EFFECTIVE THERMAL INSULATION – THE OPERATIVE FACTOR OF A PASSIVE BUILDING MODEL

Edited by **Amjad Almusaed**

Effective Thermal Insulation – The Operative Factor of a Passive Building Model
Edited by Amjad Almusaed

Published by InTech

Janeza Trdine 9, 51000 Rijeka, Croatia

Publishing Process Manager Martina Blecic

Technical Editor Teodora Smiljanic

Cover Designer InTech Design Team

Additional hard copies can be obtained from orders@intechopen.com

Effective Thermal Insulation – The Operative Factor of a Passive Building Model,
Edited by Amjad Almusaed
p. cm.
ISBN 978-953-51-0311-0

For EU product safety concerns:
IN TECH d.o.o., Prolaz Marije Krucifikse Kozulić 3, 51000 Rijeka, Croatia,
info@intechopen.com or visit our website at intechopen.com.

Meet the editor

Dr Amjad Almusaed was born on January 15, 1967. He holds a PhD degree in Architecture (Environmental Design) from "Ion Mincu" University, Bucharest, Romania. He followed a postdoctoral research in 2004 on the Sustainable andbioclimatic houses, from the School of Architecture in Aarhus, Denmark. Dr Almusaed has more than 22 years of experience in sustainable architecture with innovative orientation upon architectural science. He has carried out a great deal of research and technical survey work, and has performed several studies, in the area mentioned above. He is an active member in many international architectural associations. He published many papers, articles, researches, and books, in different languages.

Contents

Preface IX

Part 1 Passive Building Model and Thermal Insulation 1

Chapter 1 **Improvement of Thermal Insulation by Environmental Means 3**
Amjad Almusaed and Asaad Almssad

Chapter 2 **Passive and Low Energy Housing by Optimization 23**
Amjad Almusaed and Asaad Almssad

Part 2 Energy Efficiency upon Passive Building 43

Chapter 3 **Traditional Houses with Stone Walls in Temperate Climates: The Impact of Various Insulation Strategies 45**
Francesca Stazi, Fabiola Angeletti and Costanzo di Perna

Chapter 4 **High Energy Performance with Transparent (Translucent) Envelopes 61**
Luis Alonso, César Bedoya, Benito Lauret and Fernando Alonso

Chapter 5 **A New Method for Numerical Modeling of Heat Transfer in Thermal Insulations Products 81**
Sohrab Veiseh

Preface

A true architecture is where the thinking and human feelings come into play and create something completely harmonious, which ensemble structure possesses a significance. A building has to be both poem and machine. Few buildings achieve this felicitous equipoise. Those that are sensually motivating often lack sound construction technique, or fail to fully meet operational requirements. The idea of sustainability is a philosophical standpoint or a procedure that is looking for the ideal. It is based upon our looking back at the damage that our interaction with our environment has caused and determining for the future to guarantee that our actions today do not adversely influence the world for generations to come. Sustainable development has emphasis on limiting infrastructure and material usage and helping contribute to affordability during the construction of a project, by eliminating some costs altogether. In the longer term, sustainable design's principles of energy and healthy architectural spaces and material durability help to make a building affordable.

Renewing ability is the key to our human range and our prime resource for architecture. The development of the energy sector is especially relevant as it is inevitably linked to many aspects of sustainability, e.g. protection of the natural life-support systems. The first law of existing energy phenomena in ambient consider that the energy form cannot be creased or destroyed. Its form may be changed, but its magnitude persists. This is the essence of the first law of thermodynamics, more commonly known as the law of conservation of energy. The gap, which has existed in the past between the pure theory of heat transfer and its practical demand in a thermal insulation system, has been bridged. Passing from the term of thermal insulation such factor to thermal insulation such process, is something like moving from a line with one dimension to area with two dimensions. Insulation is defined as a material or combination of materials, which retard the flow of heat.

Many hypotheses, objects and decisive factors are involved in this research. The passive building provides us with the chance to reach extremely low levels of energy consumption by employing high-quality, cost-efficient measures to general building components - such measures are in turn off advantage to the ecology and economy sector. Passive building offers an exciting opportunity to achieve environmental and economic benefits. The concept of passive building deserves a deeper explanation. The hypothesis is that this affiliation leads to positive responses in terms of human performance and health even emotional states.

The new movement aims to make environmentally friendly energy upon building more effective and development process occurs by effectively managing of natural resources. This path will discover a far deeper integration of nature with the built environment and the potential synergies in exchanging energy and nutrients across the human-nature interface. Thermal insulation material is a traditional material that extends to several thousands of years, using thermal insulation in practice, from skins and mud to space, fibers and fillers. The materials can be modified to any size, form or surface. By diminishing of heat loss or gain, insulation can assist to sustain process temperature to programmed value or within a predetermined variety. The insulation thickness has to be adequate to limit the heat move in a dynamic system or diminish the temperature change, with time, in a static system. The thermal belongings of insulating materials and other standard fishing vessel building materials are identified or can be precisely calculated. The sum of heat transmission throughout any combination of materials can be calculated. On the other hand, it is essential to recognize and understand technical terms to be able to calculate heat losses and appreciate the factors that are concerned. Only in recent years has humanity's earliest accomplishment, protection from temperature extremes, become satisfactorily regarded as to constitute a ground of engineering and a science.

It is easy to determine the effective thermal insulation material fitting for different building's programs. Moreover, it will be tricky when we talk about thermal insulation systems in such a process. Today, the orientation is to establish which optimal and which competent system is to be used in different structures of buildings. It can be a phenomenon, a mathematical method, or a sustainer material. Our assignment is to find the optimal solution to be selected.

This book provides the theoretical heat transfer issues based on practical ways in order to give worked-out examples of such problems in terms of practical use.

Amjad Almusaed
Archcrea Institute, Aarhus
Denmark

Part 1

Passive Building Model and Thermal Insulation

1

Improvement of Thermal Insulation by Environmental Means

Amjad Almusaed[1] and Asaad Almssad[2]
[1]Archcrea Institute, Aarhus
[2]Umea University, Umea
[1]Denmark
[2]Sweden

1. Introduction

Insulation is a vital part of all contemporary buildings; it performs many functions, all of which influence the cost of the building and its operating cost. This component is essential to be positioned not only in the floors, walls, and ceilings of the buildings, but also using of other key technique to improve the insulating process (John F. Malloy 1969). There are many other ways, although one of the most vital way is using of vegetate buildings concept.

Building and garden usually do not arise together and seldom at the same time. Mostly a building is build first and the garden made around it, but if there was a garden first, little of it remains undisturbed by the time the building is build. The layout of the building requires a planner, that of the garden also. Often the representatives of these two entwined disciplines do not meet, or come together to late, when they can but tolerate each other. It would be better if they met to discuss and decide every detail before the first sod was broken. Best of all, the planning of building in its garden should be a mutual undertaking. (Jan Birksted 1999). No Architectural concept is complete without natural areas. Exclusive of soil such growth media to grow plants or vegetations, without water to encourage them, and without the wildlife attracted by the sustenance thus offered, an architectural element has not the fully rounded totality of a factual architecture. The most important class of environment means in reading of this chapter is green areas inside and outside the architectural elements, which requires be to implicit more in terms of ecology as an interface between us and the natural world. Therefore, a green building comes into sight such global human requirement. Today the requirement is to oriented building components towards natural resources to be included in building concept. The green areas is the most significant environmental means, where the green covering concept can be changed to the concept of biophilia.

Biophilic habitat combines the interests of sustainability, environmental consciousness, green areas of the large nature, and organic approaches to evolve design solutions from these requirements and from the characteristics of the site, its neighborhood context, and the local microclimate. The concept of biophilic architecture is a part of a new concept in architecture, that labor rigorous with human health, ecology and sustainability principles, such a integrate part of architectural configuration, which must be in optimal proportion with other buildings area. At what time an architectural element is viewed as an ecosystem, it is obvious that biophilic architecture can play a vital role in creating a healthy indoor environment.(David Pearson, 2004). The biophilic architectural concept deals with the interaction and interrelations of communities of human and plants with under architectural spaces upon local microclimate. A green areas concept can improve the building functions by increasing the efficiency of energy resource, and reducing the building impacts on human health and the environment during the building's lifecycle, through better sitting, design, construction, operation, maintenance, and removal. (Frej, Anne B. 2005).

Energy is fundamental to all life. Even early man knew his life depended upon energy from food, fire, and from the sun, and he conserved it to the best of his ability. He stored food, built a shelter around his fire, and wrapped himself in skins. The shelter around the fire to contain its heat, and the skins wrapped around his body to retard the flow of heat from his skin to the surrounding air were two types of thermal insulation. Therefore, thermal insulation was one of man's first inventions. This illustrates that the need for energy conservation is as old as man himself (John F. Malloy 1969). The new orientation of actually researches on biophilic habitat aims to move the human actions under an architectural roof towards the green of the large nature; this movement intends to create:

- Natural and physical frameworks become more than friendly.
- The Energy consummate by our buildings is most well organized.
- The human development by effectively managing of natural resources is effective.
- The negative effects of climate change become more reduced.

2. Energy consumption and macro-environment metropolitan

2.1 The negative effects of global climate change

Throughout mainly of the geological record, the Earth had been bathed in uniform warmth such was the fixed opinion of geologists. The glacial epoch it seemed to have been a relatively stable condition that lasted millions of years. During the last 2 billion years, the Earth's climate has exchanged between a frigid "Ice House", like today's world, and a sweltering "Hot House", like the world of the dinosaurs. Global climate change is reasoned by the accumulation of greenhouse gases in the lower atmosphere. The global concentration of these gases is increasing, mostly unpaid to human activities, such as the combustion of fossil fuels (which release carbon dioxide) and deforestation (because forests remove carbon from the atmosphere), cities extending and wrong consumption of our natural resources. Extreme weather events such as droughts, floods, cyclones and frosts may affect areas previously unaffected or strike with increased frequency. The sun influences Earth's climate (Amjad Almusaed 2010). What is new is that the changes predictable to occur as quickly that nature will have more than tricky to keep up. When the climate revolves out to be warmer, we have to remain for that some species will get it too hot for us, but could flourish further north (McMichael, A. J., and Haines, A. (1997)).

Human beings are exposed to climate change through changing weather patterns (temperature, precipitation, sea-level rise and more frequent extreme events) and indirectly through changes in water, air and food quality and changes in ecosystems, agriculture, industry and settlements and the economy.

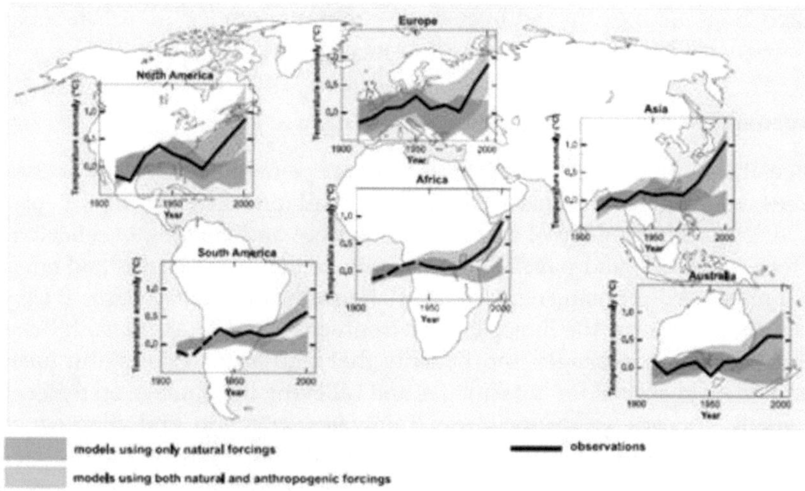

Fig. 1. The global and continental temperature change (Source: IPCC 2007)

2.2 Negative effects of urban head island phenomenon

One of the most important issues facing Biophilic cities of the future is the urban heat island effect, which will be greatly make worsted by rising global warming. The major reason of the urban heat island is change of the land surface by urban progress; waste heat creates by energy usage is a secondary contributor. As inhabitants centers grow they are inclined to adjust a greater and greater area of land and include an equivalent amplify in average temperature. Partially as a result of the urban heat island effect, monthly rainfall is about 28% greater between 30-60 kilometers downwind of cities, compared with upwind. Heat islands can affect communities by increasing summertime peak energy demand, air conditioning costs, air pollution and greenhouse gas emissions, heat-related illness and mortality, and water quality (Amjad Almusaed 2010). They can be developed on urban or rural areas. As it would be predictable, there is a minor fact regarding non-urban heat islands, since they typically do not correspond to a risk for the human being or the environment. In the meantime, urban heat

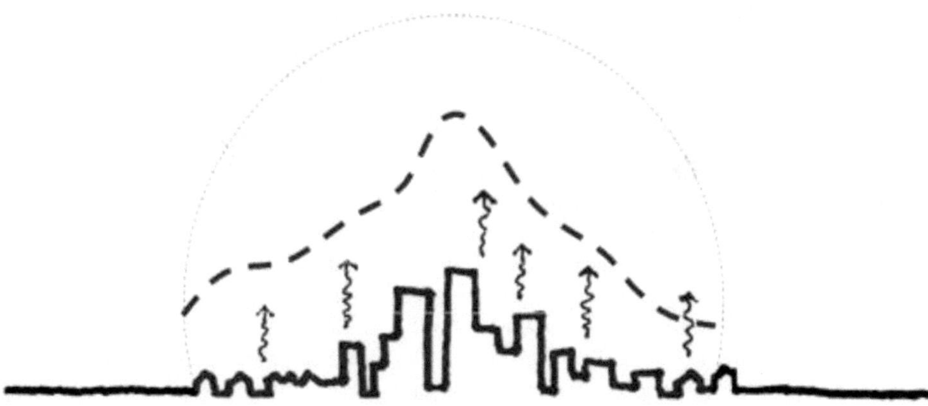

Fig. 2. Urban heat island dealings (sources: Amjad Almusaed 2010)

islands have been abundantly addressed throughout decades in urban areas with an extensive variety of climates and landscapes (Amjad Al-musaed 2007).

2.3 Improvement of energetic macro environment

Plants, vegetations upon building surfaces and are a method not only to decrease city temperatures but also to diminish the heating load and energy require of individual buildings. A long-term strategy of planting shade trees and creating of reflective buildings materials for roofs walls, and pavements can mitigate the urban heat island effect and help to diminish associated economic, environmental, and health-related costs. (H.Y. Lee 1993). Green areas supply always the important environmental and human health benefits which cover a large area of advantages and benefits that can be for example in ameliorate the urban island effect in special for hot climates and relieving the damage on the ecology of the city. Principally concerning microclimate, rainwater retention and filtering of airborne pollutant lowering energy expenditures, purifying the air, reducing storm-water runoff, longer durability of the building skin, due to lower surface temperatures and better protection against UV-radiation, creation of recreation areas in parts of the city, aesthetical improvements in denaturalized urban centers and many others. Numerous reimbursements can result from the adoption of green areas over the buildings and using the new concept of biophilic city. Vegetative building exterior skin can also play a vital function in addressing UHI in global cities, as they have been well-documented to decrease building surface temperatures and building heat gain (Liu & Baskaran 2003, Del Barrio 1998) and are rising in status due to their thermal and ecological characteristics.

One introduces additional green areas into the built environment, and the other engages choosing correct building materials that reflect the sun's rays. Both strategies diminish the urban heat island effect - the temperature in center cities is at 2-10 degrees higher than in nearby rural areas. With using of a light-colored building surfaces and materials or reflective coatings lowers surrounding temperatures. These measures may limit the frequency, duration and strength over periods of hot weather. Strategies to reduce overheating, such as the use of cold skin building and clean sidewalks, and planting trees providing shade, have many advantages.

3. Energy consumption and thermal buildings micro-environments

3.1 Reducing of energy consumption upon micro-environment by using green areas

Communities can take a many steps to save energy consumption upon micro-environment. These strategies include: By means of greater, the concept of biophilic urban and architecture, vegetated buildings extern surfaces, by living green walls and planting trees and vegetation employ the evapotranspiration and evaporative-cooling procedures of vegetation on construction surfaces and integrate open green spaces. In addition, trees, shrubs, and other plants help reduce ambient air temperatures during a process known as "evapotranspiration." This happens when water absorbed by vegetation evaporates off of the leaves and surrounding soil to naturally cool the surrounding air. Trees also insert oxygen to the atmosphere, break down a quantity of pollutants and diminish dust (Amjad Al-musaed 2007). It has been predictable that 300 trees can counterbalance the quantity of aerial pollution that a human being generates in a life span. 1 m^2 of green areas can remove up to 2 kg of airborne particulates from the air every year, depending on foliage type.

Reducing the level of heat-absorbing surfaces such as paved, asphalt or concrete surfaces and amplify their permeability, where the certain that the individual built form's configuration (size, clustering and form) does not give confidence heat-island effects (Myer, W. B., 1991).

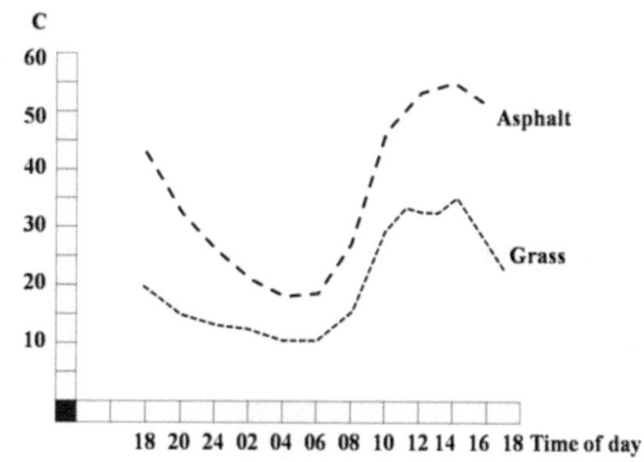

Fig. 3. Earth Surface temperature through 24 hour (Source : Amjad Almusaed 2010)

The current surfaces (roofs, infrastructure, pavements, etc) with vegetated surfaces such as green roofs or green gardens and open - network road surface or specify cool materials to decrease the heat absorption.

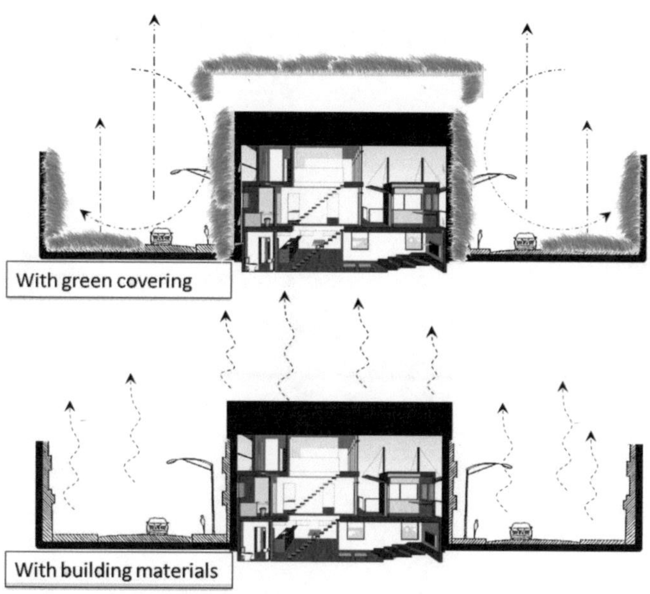

Fig. 4. The comportment of different surfaces (green covering – non green covering)

The replacement of vegetation by streets, buildings and asphalt, frequently guide to a greater absorption of sunlight throughout the day and a slow release of heat throughout the nighttimes. Selection of building material is a key in overturning the heat island effect, for it is the dense dark-colored structures that draw sunlight and keep it for periods. Green walls or roofs, with their landscaping and incorporation of natural materials, are ideal in their resistance to heat absorption. A study by Singapore researchers found that such gardens reduce roof ambient temperature by 4 °C and that heat transfer into the rooms below is lower.

A study in Tokyo shows that if the temperature in Tokyo goes down by 0.8 °C because of rooftop gardens, electric-bill savings equivalent to approximately $ 1.6 million per day could be achieved (Wong Nyuk Hien 2008). The urban heat island mitigation strategies, can support to diminish direct energy utilize in buildings, and if applied on a community-wide basis, can decrease generally ambient air temperature in a specified region (Gallo, K.P.; Tarpley, J.D. 1996).

3.2 Reducing of energy consumption by using of soft cool material buildings

Using soft cool building materials and controlled to cool paving materials. Adjust current and new urban city block layouts and configurations with explain patterns, materials and surfaces that absorb a smaller amount of solar energy.

3.3 Increasing of the shading effects

That can take place by assemblage of physical volumes, or planting trees. Planting shade trees reduces the amount of heat absorbed by buildings by directly shielding them from the sun's rays. A local microclimate can be different from its surroundings by receiving supplementary energy, consequently it is a modest warmer than its surroundings. On the other hand, if it is shaded it could be cooler on average, because it does not acquire the direct heating of the sun. Its humidity may vary; water may have accumulated there production things damper, or there may be a smaller amount water so that it is drier.

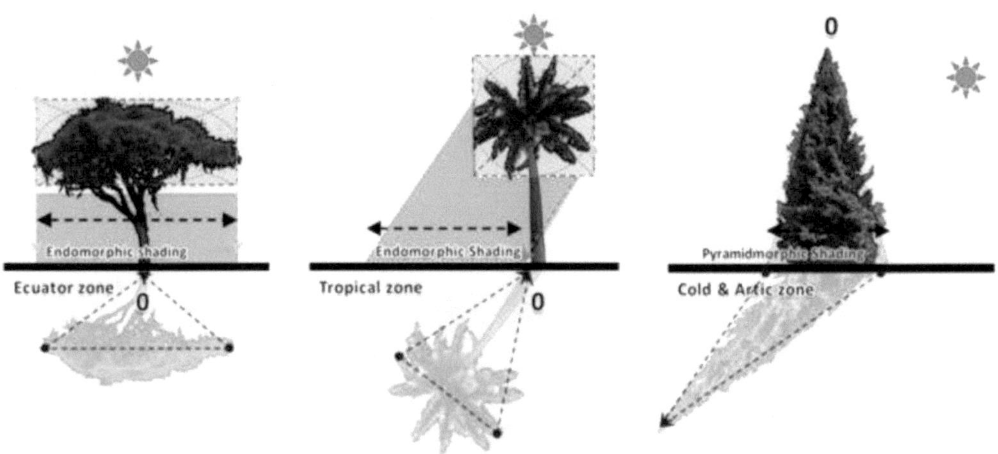

Fig. 5. Trees shade morphologic in correspondence to world climate specific (Archcrea institute)

3.4 Saving energy by using a well reflecting and high building materials

For generate a competent result, of reducing energy consumed by building function, we need to utilize a well reflective and high emissivity building materials for the climatic skin building surfaces or install green areas for the extern roof and facade. Therefore, we require to increasing the reflectivity of buildings surfaces such as rooftops and using frequently of light colors. Creation highly reflective building surfaces will keep buildings cooler and warmer and reduce energy bills. Research conducted in Florida and California indicates that buildings with highly reflective surfaces require up to 40 percent less energy for cooling than buildings covered with darker, less reflective roofs. Opt for roof, surface and building colors so as to decrease effects (evade black or dark colors but utilize white and light colors).

3.5 A well design of circulation arteries

Design the roads and street canyons width, height ratios and their orientations to control the warming up and cooling processes, the thermal and visual comfort conditions, and assist in air-pollution dispersal (Ken Yeang 2006). Design the built form with the topography of the locality, to ensure that the heat-island effect does not affect the climate of the larger region surrounding the designed system and to reduce the wider impacts on people and on the surrounding natural and built environment (Ken Yeang 2006).

3.6 Slow thermal reactions leading to formation of ozone pollution

As a result we require to control the traffic-systems reduction, distraction and rerouting to reduce the production of air pollution, and heat discharges. For parking the optimal solution is in building vehicular parking spaces underground or as covered structured parking. Use an open-grid pavement system (with impervious surfacing such as porous concrete) for the parking-lot areas (Ken Yeang 2006).

3.7 Reducing of the energy consumption

In the past, green areas on the roofs have been used to insulate edifices. The major and vital role of green areas on biophilic architecture is that to conserve, insulate and hold back a change of energy flux, between outside and inside. The green areas amplified the thermal performance of the green covering system and constantly lowered the heat transfer between the construction and its environment all over. Green areas insulate buildings by preventing heat from moving throughout the climatic skin areas.

3.8 Increase of the physical comfort and the quality of the life

The economic price of the success strategies is outweighed not merely by the cooling energy reserves, but in addition by the decrease in greenhouse gas releases, esthetic value of urban forestry, and the increased quality of human health (Hinkel, Kenneth M. (March 2003). These can be defined as win strategies. Mitigation of the urban heat island impact by increasing the employ of surfaces covered in vegetation and building materials with higher than usual reflectivity; in mixture with a strengthening of emissions decreases programs has the potential. Using top roof such climatic skin roof can help our mitigation strategy for reducing of urban heat island effect.

3.9 Reducing of the buildings height

Using of a very high buildings in the centre of the cities increase temperature few degrees. The high buildings surrounded by many urban areas give a multiple surfaces for reflection and absorption of sunlight, increasing the efficiency with which urban areas are heated. This is called the" canyon effect"

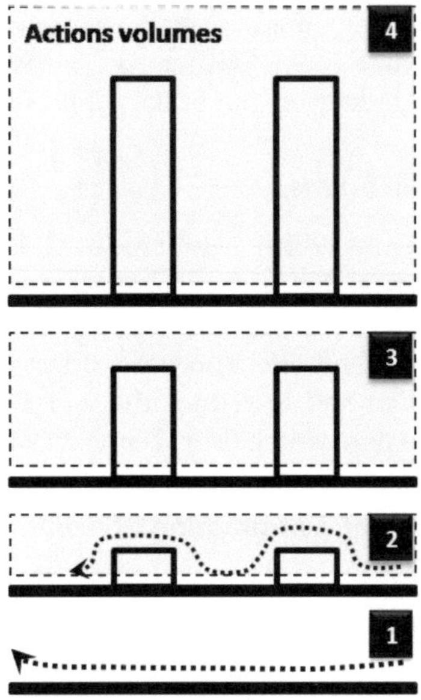

Fig. 6. The effects of high building on city climate

4. Thermal performance of green covering

Green areas are the most important visual associations between land, buildings and the sky; the most prominent of all plant life, and without their presence, our townscapes would be naked. A sense of continuity is given by old green area and they remain well-known marker when unneeded buildings, hedgerows and path make way for new developments. The green areas amplified the thermal performance of the green covering system and constantly lowered the heat transfer between the construction and its environment all over. It insulates building by preventing heat from moving throughout the climatic skin areas. For cold and temperate climate the energy flux occurs from hot inside spaces to cold outside environment and contrary meant for hot climate. Thermal insulating green area build up with official property values are permitted to be supplementary to the conventional thermal insulation.

Due to this special build-up, the building owner saves approx. 2 litters / m^2 fuel oil per year. The green areas on building surfaces reduced the daily energy demand due to heat flow through the building surfaces by 83-85 % in the spring/summer and 40-44 % in the fall/winter, with an overall annual reduction of 66 %. Green areas insulate buildings by preventing heat from moving through the climatic skin areas. Their insulation properties can be maximized by using an increasing medium with a low soil density and high moisture comfortable and by selecting plants with a high leaf area directory. In the winter, the additional insulation supplied by the growing medium (substrate) helps to diminish the amount of energy necessary to heat the building. The amount of the energy rate savings impact is a function of (Amjad Almusaed 2008):

- The size of the building

- The building location
- The depth of the growing medium
- The type of plants and other variables

Since the 1980s investigate has been conducted on topics such as the insulating effects of greens on façades. Green areas over building surfaces have been shown to significantly reduce building surfaces temperatures and building surfaces heat gains. Karen Liu's field experiments in Ottawa, Canada confirmed that an extensive green area reduced heat gains by 95 percent and reduced heat losses by 26 percent as compared to a standard reference area (Liu & Baskaran 2003). In experiments at Pennsylvania State University (PSU), roof surface temperatures were below ambient air temperatures in greened roof areas at the same times that temperatures on traditional roof surfaces reached 40 degrees Celsius above air temperature. PSU studies also indicated significant (5 - 10 degree C) differences in indoor air temperature in rooms below greened and non-greened roof areas (Gaffin et al 2005).

The vegetative skin building was modeled as three divide layers - the building material layer, the soil surface layer, and the canopy layer. Each is represented by its own energy balance, as seen in the figure below. The associated equations can be linked together by the flux through each connecting boundary. This makes it possible to solve for surface temperature taken at the soil surface, as it is most easily measured in a green roof. Heat flux into the building can also be solved for using these energy balances, as well as water lost through evapotranspiration.

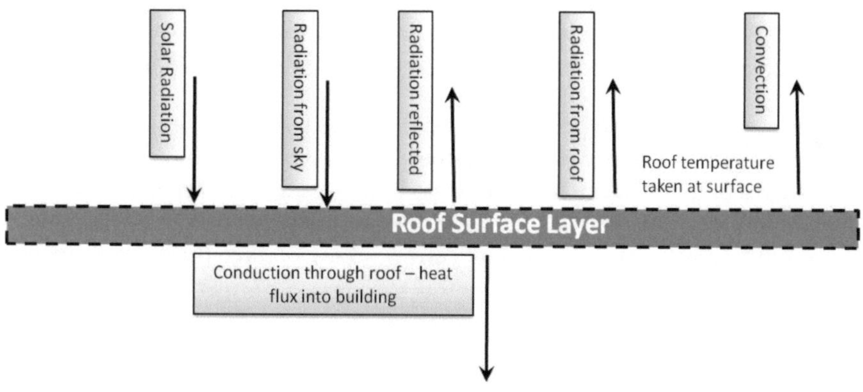

Fig. 7. Energy balance on roof surface layer for typical, reflective, and elastomeric roofing (Source : Caroline H. 2008)

Temperatures of classical building surfaces exceeded ambient temperatures by up to 45 °C and had ranges of skin building temperature also exceeded 45 °C. Vegetative skin buildings maintained temperatures under the ambient during the day in the majority cases, and had average temperature levels that were at or under the ambient environmental temperature (Caroline H. 2008).

The average range in temperature for a green skin building was 10 °C, while the average range in classical skin building was 42 °C. These vast roof temperature ranges can create stress in the structural roofing materials themselves, which is one skin building of the reasons that green skin buildings are able to extend the life of building materials. Vegetative skin building temperatures and fluxes were the lowest category of skinning building condition during the midday hours in the greater part of cases, declining midday temperatures and fluxes would be mainly significant in office buildings that not only have highest solar heating loads at that time, but also highest heat loads from high habitation and equipment operation.

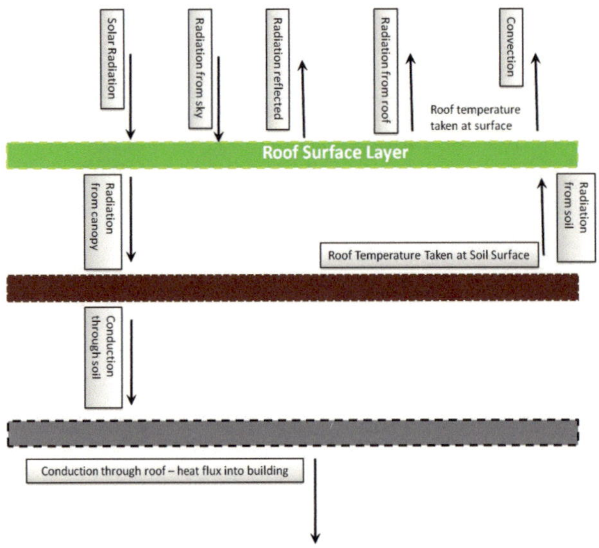

Fig. 8. Energy balance on vegetative skin building layers (Source : Caroline H. 2008)

Additionally, the temperature of green skin building surfaces and the heat flux through the green skin buildings had a lower range of values than any other kind of skin building. While the average values of temperature and flux were often inferior for elastomeric skin buildings, the range of values was much lower for vegetative skin buildings. The elastomeric building kin surfaces much more regularly had negative values of flux (heat loss) mainly in the morning and evening hours. These negative flux values of elastomeric skin buildings were in several cases as large as or larger than the maximum positive flux. The fact that green skin building temperatures and flux values were most stable throughout the day, representing the lowest range from morning to midday, is also considerable for building operators to note. This means that cooling and heating loads will be consistent throughout the day. While any kind of alternative skin building was confirmed to greatly decrease the flux into the building, it should also be an objective to maintain flux and temperature so as to decrease the pressure put on heating and cooling systems to adjust for changing heat fluxes. In particular, the negative heat fluxes that often were demonstrated by elastomeric skin buildings in the model results would indicate that buildings might need heating in the mornings to maintain room temperature due to this heat loss(Caroline H. 2008).

4.1 Greenly areas placement

The green areas can take a differ places in relation to the non- greenly areas where the green area appearance aim to be synchronized by means of other area in concordance with architectural perception upon biophilic habitat. The stabilizer forms resulting from the accumulation of separate elements, which can be characterized by their capability to develop and combine with other forms. For recognize preservative groupings as integrated compositions of shapes as figures in the visual field, the combining elements have to be connected to one another nr a rational method. Good biophilic habitats plan their planting to avoid unfavorable local microclimates avoiding frost pockets for sensitive crops, and allowing for the effect of aspect on temperature or water balance. They can also try to make new microclimates, which will favor the plants they are growing. Shelterbelts of planted trees or bushes create a drag that slows down the drying or cooling winds that blow across architectural volume. The effect of a shelter belt of trees on wind speed can extend across the field as far as 20 to 30 times the height of the plants (Jonathan Adams 2007). By means of the green areas form and position over architectural concept, it can be measured by three criteria:

- Performance.
- Identity.
- Economy of means.

Everyone has a subconscious or usual means to be familiar with the architectural elements that are used every day symbols of comfort, familiar functions and occasionally, visual excitement (Amjad Almusaed 2010).

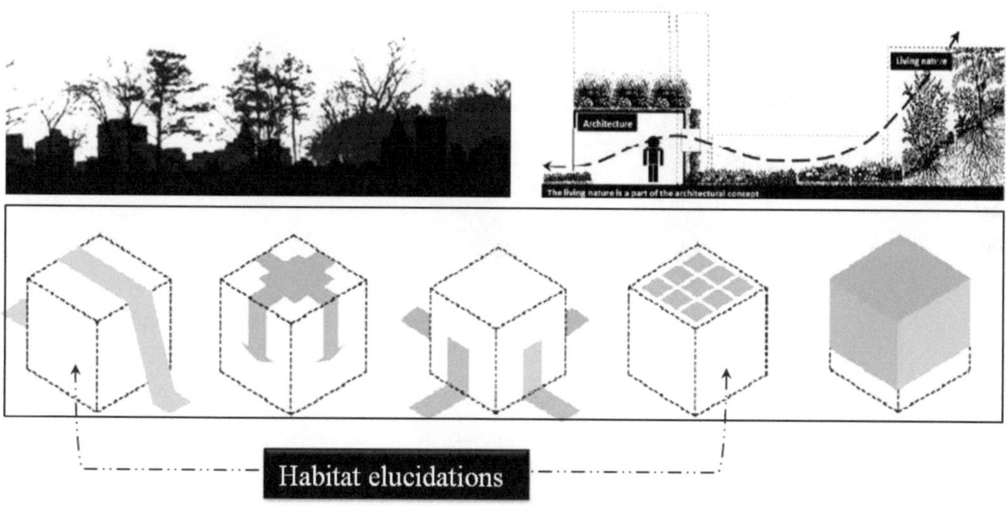

Fig. 9. Architecture and green covering forms and placements upon biophilic habitat (Amjad Almusaed 2010)

4.2 Greenly effects on the environment

The carbon is incorporated into the tree's growth. Because of transpiration and shading, the air surrounding a tree can be around 5 °C cooler than its environment. Tree-shaded neighborhoods can be up to 3.5 °C cooler than those without trees. The competence of plants to produce oxygen varies quite a bit. It is also potential to build an artificial process involving photosynthesis that would successfully do the same thing but it would not be a beautiful to walk through (Ken Yeang 2006). An average of human requires are; 2600 grams of food, 686 grams of oxygen (O2) and 400 grams of water.

4.2.1 Vertical green

A plant leaf produces about 0.005 litter's oxygen per hour. Therefore a mature human need about 50/0.005 = 10000 leaves which would be provided by about 500 small plants for one person.

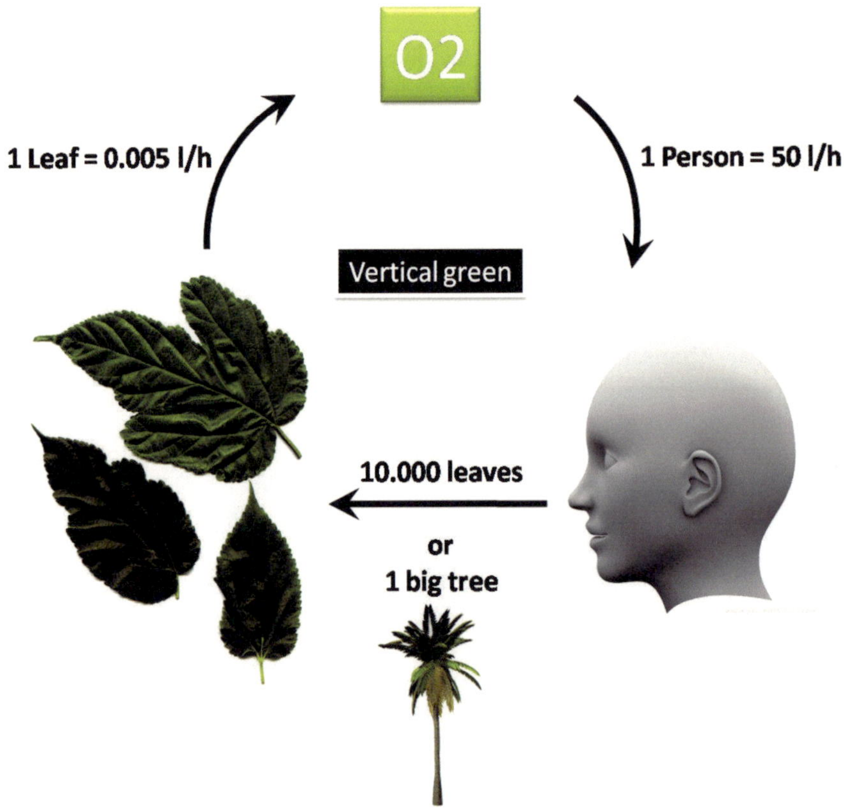

Fig. 10. Human and plant interaction on vertical green

If the average of shrub or other medium size plant has 30 leaves per plant, then that would be 5 ml / leaf x 30 leaves = 150 plants (Wizkid 2008). An average of 18 $cm^{2,}$ leaf area can release atmosphere of 0.005 litters' oxygen per hour. An average of person who consumes 50 litters oxygen per hour. Consequently, an average of 18 m^2 of vertical green area is sufficient for one person. In addition, an average of 5 m^2 of vertical green areas is satisfactory. There are many assumptions, average leaf, and average plant.

4.2.2 Horizontal green

In a 1.5 m^2 of uncut grass, produces enough oxygen per year to supply one person with their yearly oxygen intake requirement (Brian Burton 2009).

In addition it will necessitate to take into evidence oxygen production reduces as carbon dioxide concentration increases, assuming this hypothetical person is in a limited space with all these plants, the CO2 concentration will increase suitable to the person's expiration. This will slow down the plant's photosynthetic rate (Jonathan Adams 2007).

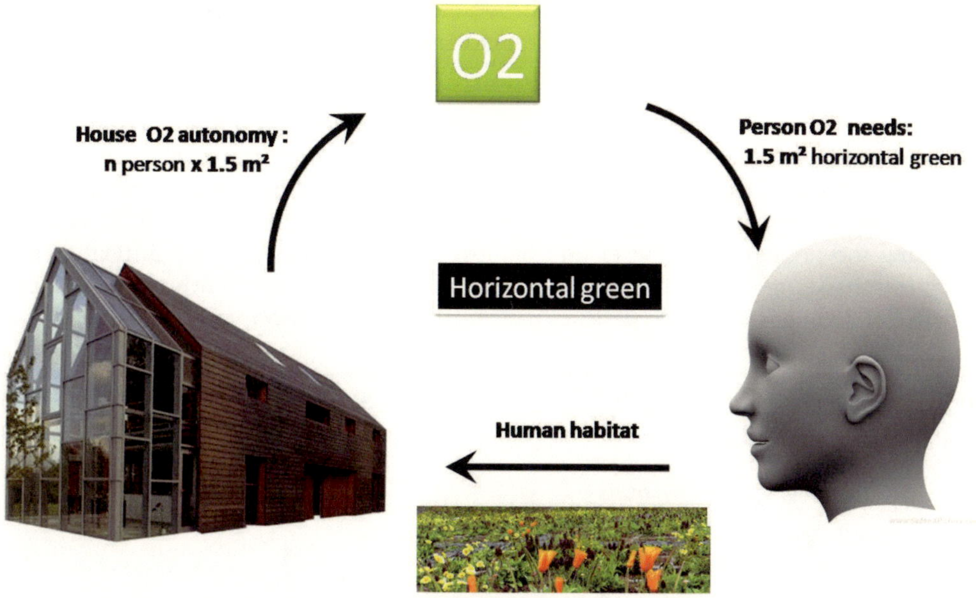

Fig. 11. Human and plant interaction on horizontal green

Hospitals and health facilities utilize the therapeutic benefits of green areas. These facilities sometimes use gardening as a tool to enhance the healing process for patients. In addition, the person can enjoy the comfort, fresh air, and landscape while restoring their health (Ismail Said (Jun 2003)). The query is how we can obtain the oxygen and air quality from the plants. biophilic structure on the earth is a valued and appreciated part of life, where areas and human carrier green is not only an excellent synthesis of both qualitative and quantitative research that documents the bond between people and plants, it is a synthesis of the life's work and thinking of one of the most important figures in people-plant relationships.

Using of a good managed green covering. According to the NASA study, the heat island effect in urban areas can be most effectively reduced with more green space (vegetation offers moisture to cool the air). In adding, light-colored surfaces can reflect sunlight, and should be used on rooftops (J. Hansen, R. et al (2001). Excessive using of solid elements with less thermal properties such as some of building materials in the front of a less using of soft materials with high thermal proprieties such plants amplify the phenomenon radically (Henry J, Glynn, Heinke G 1989).

When green areas are replaced by asphalt and concrete upon roads, buildings, and other structures, it becomes essential to provide accommodation-growing populations. These surfaces absorb - rather than reflect - the sun's heat, causing surface temperatures and overall ambient temperatures to increase see table 1.

The climate performance of the biophilic architecture be able to be considerably affected by green walls, as well the visible changes concerning temperature, the solar gain by direct solar radiation and long-wave heat as well as convection. In addition the, changes in the humidity levels are also supposed. It is essential to image what an exterior living wall will look like during the winter. Green roofs have extended term experimental in value, variety, conception (Velazquez L, Kiers H 2007). All elements can be given differently into all numbers of permutations combinations of solid and almost transparent membranes (Amjad Almusaed 2004).

Green areas are the most important visual associations between land, buildings and the sky, the most prominent of all plant life, and without their presence our townscapes would be naked. Our modern lives seem to be dominated by conflicts of one kind or another, and on the particular subject of trees it is the pressure on land and the rise of consumer power that is placing the professional adviser and his love of green areas in some difficulty. Today modern architecture may fairly be said to have won its first battles all over the world, but in very few of them has it had any assistance from landscape architecture.

Criterions	**Extensive green walls**		**Intensive green walls**		**Green roof**
	Spots green suspended walls	*Compact green suspended walls*	*Living walls*	*Energetic biophilic walls*	
	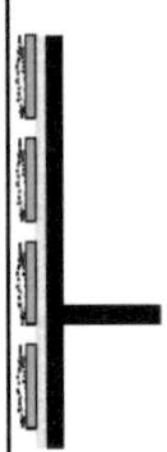		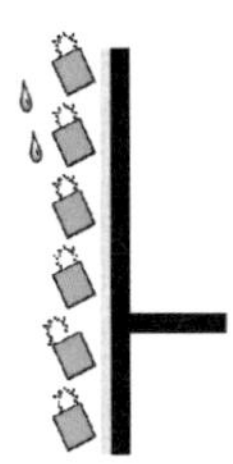	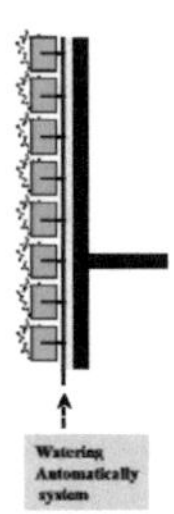	
Thermal insulation	Low, very diminutive	Low, diminutive	Middle, can be efficient	High, more efficient	High, excellent
Environment friendly	Low, is more decorative	Middle, need relationship between building and green areas	Middle, need relationship between building and green areas	High, green is a part of buildings model	High, green is a part of buildings model
Urban heat effects reduction	Low, employment	Middle, employment	Middle, employment	High, effective in combating the phenomenon act	High, effective in combating the phenomenon act

Table 1. The different categories of green walls and roof (Archcrea institute)

One of the best qualities of the modern movement is its increasing awareness of the connection between the space within building and the space around them, and of the interdependence of building and green areas.

5. Other environmental mains of thermal insulating

5.1 Double skin façade (the energetic role of double skin façade)

Walls must give building spaces protection against hot, cold, wind, external noise, and enhance security. A well insulate heavy construction is needed. But also a sustainable external element is necessary. For an efficient bioclimatic building architect can oriented to a curtain wall such as sustainable exterior elements. Curtain wall is synonyms: double skin façade double-leaf façade, double façade, double envelope, wall filter façade, and ventilated façade (Amjad Almusaed 2010).

The curtain wall on façade is principally a couple skins separated by an "air corridor". The main layer of skins is usually insulating. The air space between the skins layer is as insulation against temperature extremes, winds, and the sound. If there are two skins of glass, or other thermal opaque materials so for shading interior space that the sun-shading devices are often located between the

The double skin façade consists of two layers of materials, with air space between the two layers preserved, the principal's roles of curtain walls are controlling of solar gain, access to fresh air, embodied energy, esthetics. see figure 12.

Of course, there is a certain level of energy consumption, but this is significantly reduced as internal temperatures, are already lower than outside temperatures. Double -skin facades offer a protected from the exterior environmental conditions; these shading devices are less expensive than system mounted on the exterior.

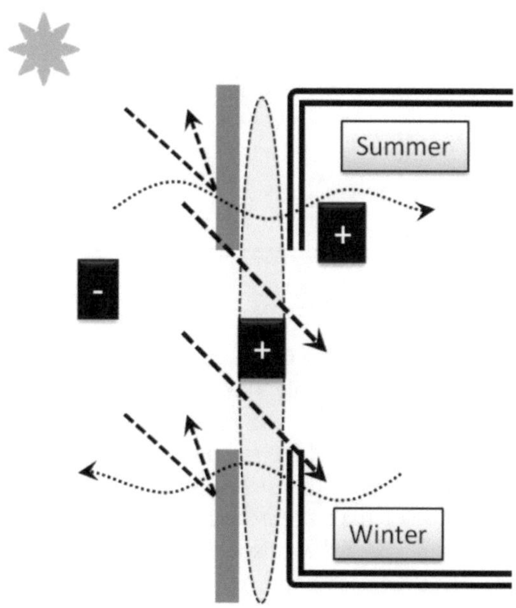

Fig. 12. The thermal role of curtain wall

The principal benefit of double-skin façades over traditional architecture is that they permit the application of blinds even for the buildings with substantial wind. The special materials are mainly used for architectural purposes due to their performance in reducing solar heat.

- Energy saving is an essential factor to reduce the emission of carbon dioxide which is a cause of global warming, and ventilation is only unique method to control the indoor air quality and also regarded as an effective method to sweep out the indoor.
- In one recently experiment, achieved by department of civil engineering and architecture in University of the Ryukus in Japan, by a group of engineers and architects, about the ventilation of living spaces. That was by using two types of passive cooling (one is a conventional building with cross-ventilation through open windows, and the other is a nuilding which has double skin walls and the gaps of these skins are ventilated .The ventilation airflow comes from the gaps between the double skins and is discharged from an exhaust tower on the roof top. It called ventilation model. The latter is an ordinary hours. The outer walls are made from reinforced concrete and the inner walls are made from heat insulation boards, whose material is foamed polystyrene, and plywood. The difference between two experiment models is only the gap between the outer and the inner skins. The resulted of this experiments shows that the indoor air temperature in the ventilation model is lower than the conventional model by 2 degrees in summer and 1 degree in winter except 5 hours in the morning (Oesterle, Lieb, Lutz, Heusler 1999).

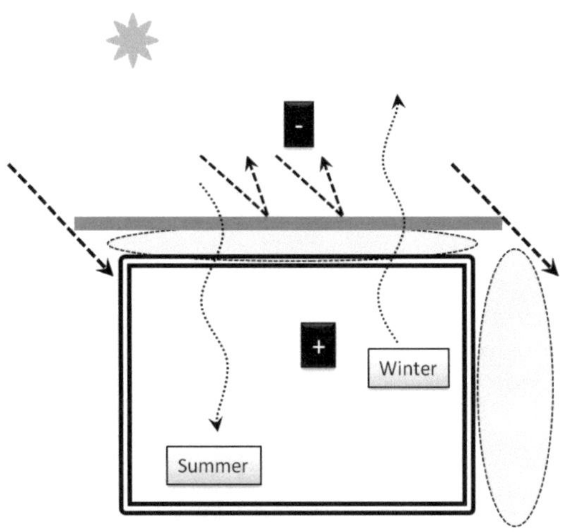

Fig. 13. The double skin roof positive effect

It is clear that the double skin walls of the ventilation model have effects on indoor thermal environment, which means that the double skin walls can keep the indoor space a little cooler all the year round. Several hours in the morning, this relation reverses, because the outdoor air temperature rises quickly and the ventilation model is directly influenced by the outdoor air. It is vital to make clear that the double skin wall can only cool the air inside the structure by several degrees lower than the actual external temperature (Amjad Almusaed 2004). It would be utopian to expect that these systems could provide the same cooling action as air conditioners. On the other hand, integration this construction form into a big intelligent bioclimatic system can give us a better resultant, which we will see that in other parts of research.

The roof on the hot climate houses receives the highest proportion of solar radiation and is also the surface barest to the clear cold night sky. To limit the heat gain, the most effective method is to shade or construct a second roof over the first see figure.

The outer roof will reach a high temperature and it is therefore imperative to separate it from the inner roof, to provide for the dispersion of heat from the space between the tow and to use a reflective surface on them both. The surface of the lower roof should reflect the low temperature heat and for the outer roof a white surface is best.

5.2 Heat break transfer concept

We all depend on energy to get better our lives. But using energy means nothing on its own; it just a way to achieve something else. And we are becoming more aware of some of the problems that come from wasting energy. The significant way of a wasting energy is energy losses by exchange of energy through external elements. Energy losses in a building mainly occur by conduction through external surfaces radiation, and convection. Conduction takes place when a temperature gradient exists in a solid medium, such external wall, windows, roofs, floors. Energy is transferred from the more energetic to the less energetic molecules when neighboring molecules in collide. Conductive heat flow must occur in the direction of decreasing temperature because higher temperatures are associated with higher molecular energies. Heat transfer through radiation takes place in the form of electromagnetic waves, mainly in the infrared region. The radiation is emitted by some body as a consequence of the thermal agitation of its composing molecules. In a first approach the radiation is described for the case that emitting body is a so called black body. Heat energy transferred between a surface and a moving fluid at different temperatures is known as convection. Condensation on the windows may be a sign of heat loss.

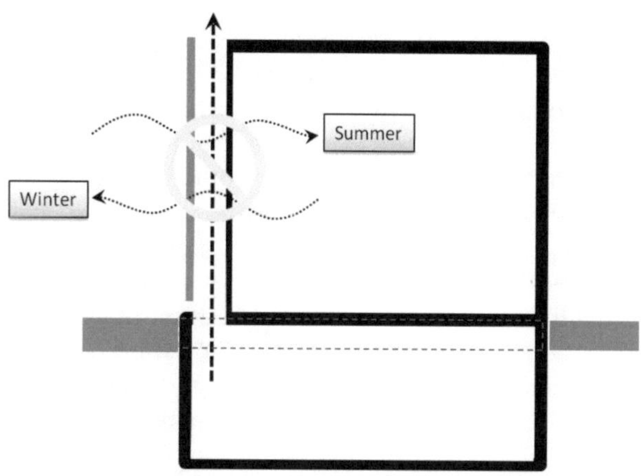

Fig. 14. The constructive effect of the optimistic underground temperature

A damp area around the window from the exterior is another sign of heat loss. During the winter, a typical window loses up to 10 times more heat than an equivalent area of an outside wall or roof. Windows can account for up to 30 % of the heat loss from a conventional house, adding significantly to heat cost. Drafts, window condensation and mould can also affect our comfort and indoor air quality. Sustainability is a wise approach to the way we live. And using energy in a more sustainable way is a part of this approach (Amjad Almusaed 2004, p187). We can save money, reduce imports, protect the environment, and move society forward in an intelligent manner. If we start doing this now we win as individuals and we will win as a society. The concept of intelligent energy losses break consists of a using of some thermal effects to wipe out or stopping the immigration of energy between exterior spaces and interior through external element. This concept can be useful using in architecture on the extreme climate regions. By a deep study of specialists about the optimal thermal effect that can help in realization this concept, consequently we must seek for a suitable source of energy, which must be permanent and easy to get. Creation of this system subsequent to the passive and zero energy concept need a well integrates of the energy in the house's components (Amjad Almusaed 2004).

6. Conclusion

A green building is a confusing expression of biophilic architecture. Green building is a construction, which can be shaped by mains of renovation process. While, a biophilic architecture strugglers the negative effects of urban heat island in local microclimate scale, and improves the human physical comfort to create a healthy human life. Therefore, one of the major problems facing us is how to establish and maintain environments that support human health and at the same time are ecologically sustainable. Green areas seems too important to people. Most people today believe that the green world is beautiful.

In fact, green areas by now contribute, some extent, to a better microclimate through evaporation, filtering of dust from the air and reduce in temperatures at the buildings surface. Besides improving the microclimate and the indoor climate, the retention of rainwater is another important advantage. Aesthetic form require, escalating the value of the possessions and the marketability of the building as a complete, mainly for accessible green areas.

On arid climates when the sun rises up, buildings roofs and asphalt road surface temperatures can rise up to 30–45 °C hotter than the air, while shaded or moist surfaces frequently in more rural environs remain close to air temperatures. City surfaces with plants offer high moisture levels that cool the air when the moisture evaporates from soil and plants (Parker, David E. 2004). The influence of plants must employment eventually to keep up with the increased require in energy. Improving energy efficiency can decrease the global warming effects of carbon in the atmosphere, improving air and water quality, and encouraging sustainable development in the cities. The physical frameworks of the city extends unprompted; consequently, it turns out to be a major area of the city centre. The green areas diminutive and take a negligible part of the city typically marginal. Many fixed edifices (civil and industrial buildings) and mobile elements such as cars, public transport and another feature contributing to the warm cities that will increase the phenomenon dramatically.

One of the most significant subjects for our study is to show how we can discover the best possible manner to realize our earth greener, sustainable, and our buildings agreeable and saves more energy, to help the human to live in healthy and economically framework.

7. References

Amjad Almusaed 2010, Biophilic and bioclimatic Architecture, Analytical therapy for the next generation of passive sustainable architecture, Springer Verlag London, UK

Amjad Almusaed 2008, Towards a zero energy house strategy fitting for south Iraq climate PLEA 2008 - 25th Conference on Passive and Low Energy Architecture, Dublin, 22nd to 24th October 2008

Amjad Almusaed 2007. Heat Island Effects upon the Human Life on the City of Basrah, Building low energy cooling and advanced ventilation technologies the 21st century, PALENC 2007, The 28th AIVC Conference, Crete island , Greece.

Amjad Almusaed 2004. Intelligent sustainable strategies upon passive bioclimatic houses, The architect school of architecture in Aarhus, Denmark. Pp203-230.

Berdahl P. and S. Bretz. 1997. Preliminary survey of the solar reflectance of cool roofing materials. Energy and Buildings 25:149-158

Brian Burton 2009, Green Roofs and Brighter Futures, http://www.newcolonist.com/greenroofs.htmlBowler P.J. 2003. Evolution: the history of an idea. California. p10

Caroline H. 2008, Modeling thermal performance of green roofs, Ecocity World Summit 2008 Proceedings, Yale College

David Pearson, 2004. The Gaia natural house book, creating a healthy and ecologically sound home Gaia books limited, UK.

Del Barrio, EP, 1998, Analysis of the green roofs cooling potential in buildings, Energy and Buildings 27(2), pp. 179-193.

Frej, Anne B, 2005. Green Office Buildings: A Practical Guide to Development. Washington, D.C.: ULI--The Urban Land Institute.

Gaffin, S, Et al 2006, 'Quantifying evaporative cooling from green roofs and comparison to other land surfaces.' Proceedings of the 4th annual Greening Rooftops for Sustainable Communities Conference. 11-12 May 2006, Boston

Gallo, K.P.; Tarpley, J.D. 1996.The comparison of vegetation index and surface temperature composites of urban heat-island analysis. Int. J. Remote Sens.17, 3071-3076.

H.-Y. Lee 1993. "An application of NOAA AVHRR thermal data to the study or urban heat islands". Atmospheric Environment 27B

Henry J, Glynn, Heinke G 1989. Environmental Science and Engineering. Prentice Hall, Eaglewood Cliffs, N. J. 07632.

Hinkel, Kenneth M. (March 2003), "Barrow Urban Heat Island Study". Department of Geography, University of Cincinnati.

IPCC, 2007: Climate Change 2007: The Physical Science Basis. Contribution of Working Group I to the Fourth Assessment Report of the Intergovernmental Panel on Climate Change [Solomon, S., D. Qin, M. Manning (eds.)].

Ismail Said (Jun 2003), Therapeutic effects of garden: preference of ill children towards garden over ward in Malaysian hospital environment, Universiti Teknologi Malaysia, Jurnal Teknologi, 38(B) Jun. 2003: 55–68

J. Hansen, R. et al (2001), A closer look at United States and global surface temperature change. J. Geophys. Res., 106, 23947-23963

Jan Birksted, relation architecture to landscape, E& FN SPON, 1999, London England. Pg 179

John F. Malloy 1969, Thermal insulation, Van Nostrand-Reinhold, the University of Michigan, USA

Jonathan Adams 2007, Vegetation-Climate Interaction, Springer in association with Praxis Publication, 2007, New Gersy, USA

Ken Yeang, 2006 A manual for ecological design, Wiley - Academy, UK

Liu, K and B Baskaran2003,.Thermal performance of green roofs through field evaluation. Proceedings for the First

McMichael, A. J., and Haines, A. (1997). "Global Climate Change: The Potential Effects onHealth." British Medical Journal 315.

Myer, W. B., 1991, Urban heat island and urban health: Early American perspective, Professional Geographer, 43 No. 1, 38-48. North American Green Roof Infrastructure Conference, pp 1-10

Oesterle, Lieb, Lutz, Heusler 1999, Double-skin façade, Integrated planning, Prestel, 87-91Olivia Nugent (April 2004), Primer on Climate Change and Human Health,edited by Randee Holmes

Velazquez L, Kiers H, 2007. Hot Trends in Design: Chic Sustainability, unique driving factors & boutique Green roofs. Proc. 5th Annual Greening rooftops for Sustainable Communities Conference, Minneapolis

Weng, Q.; Yang, S. 2004. Managing the adverse thermal effects of urban development in a densely populated Chinese city. J. Environ. Manage.70

Wizkid 2008, Plants making oxygen, USA state, energy department, Biology Archive http://www.newton.dep.anl.gov/newton/askasci/1993/biology/bio027.htm

Wong Nyuk Hien 2008, Urban Heat Island Effect: Sinking the Heat, innovation the magazine of research and technology, vol. 9, Nr.1.

2

Passive and Low Energy Housing by Optimization

Amjad Almusaed[1] and Asaad Almssad[2]
[1]*Archcrea Institute, Aarhus*
[2]*Umea University, Umea*
[1]*Denmark*
[2]*Sweden*

1. Introduction

The house is not only a roof, but also a home, the place where it is formed the moral climate and on which lasts the family spirit. UN has classified habitat settlements and identified 10 general functions that every habitat should have. (Recreations and interpretation, Preparing the foods, Eating, Relaxing and Sleeping, Study ,WC, Hygienic necessities, Cleaning, Circulation and storage, Exterior circumstances). Housing is a human right is a multi-platform documentary portrait of the struggle for home. The house, being a product of the human work, a long time user product, like any other product it has not only to be produced but also to get the user's disposal. A house is a home, shelter, building or structure that is dwelling or place for habitation by human being. Sustainable design's principles of energy and healthy architectural spaces and material durability help make a home affordable. Presently becomes incorrect work manner when we take the building phenomenon such as (passive and low energy building), detached from the large concept of architecture. (Amjad Almusaed 2004). The passive and low energy housing represents one of the most consistent concepts of sustainable building and brings with consideration of energy saving concept. Presently becomes incorrect work manner when we take the building phenomenon such as (passive and low energy building), detached from the large concept of architecture. The architectural product, being a product of the human work, a long time user product, like any other product it has not only to be produced but also to get the user's disposal. The human comfort is a vital aim of architecture, and it classified such variable level. The interaction always appears between the energy such abstract act and human comfort such human feeling. The balancing condition is extremely complex.

2. Between architectural and building concepts

Sustainable building design involves a wide range of complex issues within fields of building physics, environmental sciences, architecture and marketing. Sustainable building design views the individual building systems not as isolated entities, but as closely connected and interacting with the rest of building and a large sphere of environments (Robert Hastings & Maria Wall 2009). The Passive and low energy house idea is both easy and very tricky. It represents one of the most consistent concepts of sustainable building and brings with consideration of energy saving concept. Sustainable architecture is more than energy efficient or zero-emission building. It must adapt to and respect its environment in the broader context of "milieu". This encompasses the natural, ecological, bio-economic, cultural and social setting (Robert Hastings & Maria Wall 2009). A high quality of sustainable building brings comfort primarily up-to-date and durable products to the building user with lowest current energy costs.

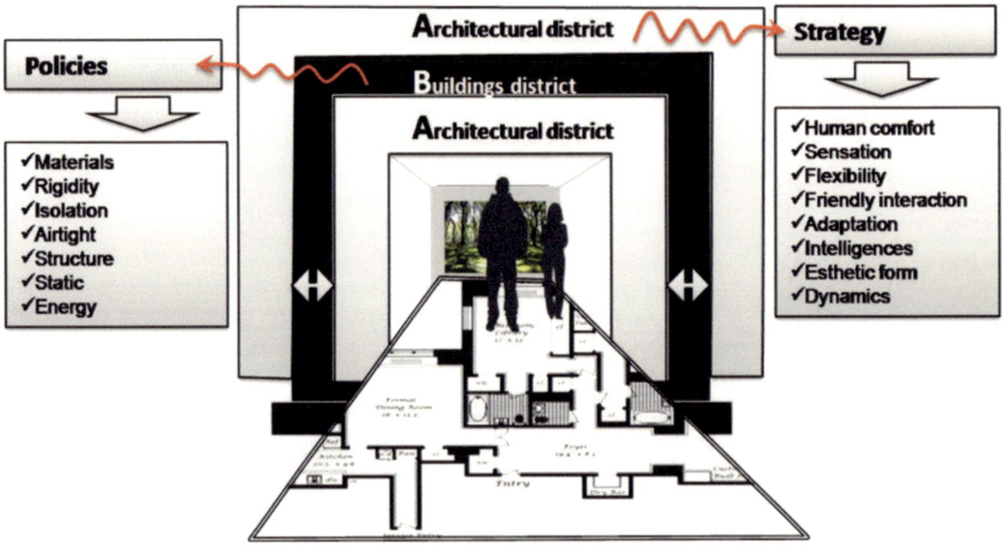

Fig. 1. Building and architectural concepts differences

As soon as we talk about passive and low energy housing, many suppose that we talk about a machinery-house concept, a building without human sentiment. Others believe that passive and low energy housing is an ugly creature. A lot of engineers, designers, agriculturists, etc. wrote about sustainable, passive or low energy buildings, green buildings, etc. Although a small, part of them reached the right concept of passive and low energy housing in concordance with architectural theory. Therefore, we can identify the technical nature of these concepts written by them. It is a big difference between the term of "Building" such a policy and the term of "Architecture" such strategy. "Building and its component" is a policy on human design, which accepts the terms of passive and low energy concepts, while "Architecture" is a strategy, which includes a large diversion of policies (Amjad Almusaed 2010).

3. The reasearch area

The main object of this research is to build a housing strategy, which integrated the concept of passive and low energy building in architectural theory; this can be occurred by generate a measurable architectural concept that includes all variables and constants factors. The interaction is between the house affordable concept, passing through a maximum healthy, comfort and esthetically along with a less uses of energy and then more economically.

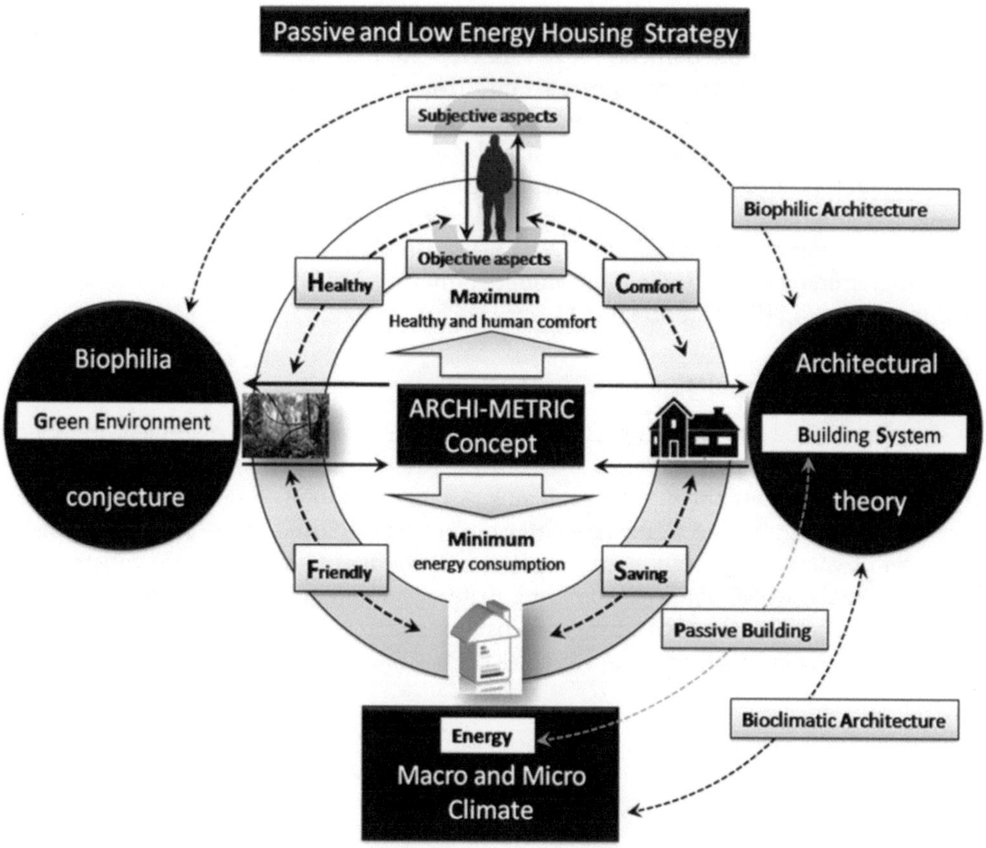

Fig. 2. The interactions of factors which intervene in passive and low energy strategy

Presently becomes incorrect work manner when we take the building phenomenon such as (passive and low energy housing), detached from the large concept of architecture. In our research we need to build a selective conception of housing, where all factors takes in evidence, environment, biophilia, energy and microclimate.

4. Invistigations method

Investigate will lucid two means:

4.1 Archi-Metric method

One of the significant objectives of optimized concept is to create a balancing system of a large size of factors, elements and concepts. We have to labor with energy efficiency, human comfort and economy, which, provides us with the opportunity to reach extremely low levels of energy consumption by employing high quality, cost-efficient measures to general building components. Our assignment is to repatriate human requirements on buildings ability, by means of maximum advantage, minimum disadvantage and optimal solution. "Archi-Metrics" method is a model, which, aids in converse all architectural phenomenon topics to be measurable with numerical characters, by using of mathematical models of

"Operation Research" science. The "Operation Research" is a mathematical method that transforms the human phenomenon and behaviors to logic mathematical models. It labor with the maxim advantage, which can acquire from different variants, and the minimum disadvantage of a negative environment action resulting from factors. At last we have to find the optimal solution between many variables and constant such an intersection point of a many variables curves. Several algorithms are available which can be used for the method of nonlinear programming problems. The problem is a nonlinear optimization problem with nonlinear constraints and cannot be solved using standard optimization methods such as linear programming or quadratic programming. Improved move limit method of Sequential linear programming (Rekha Bhowmik (4. april 2008)).

4.1.1 The problem formulation by LP technique

The simplex method for resolving Linear Programming problems is extremely influential. Therefore, a number of techniques for resolving nonlinear programming problems are supported by converting them to LP problems. An initial solution is to be selected to supply a base for the determination of the tangents of the constraints and of the objective. Consider a finite set of variables $x1, x2, \ldots, xn$. The unit cost coefficients for the main constructional elements, namely, walls, windows, building materials, thermal insulation, etc, are assumed and the construction, negative acting or cost function, $f(x)$ is to be minimized. This is generally a nonlinear function of the variables (Rekha Bhowmik (4. april 2008)).

Thus, the problem is:

minimize $f(x)$

subject to $g\,j(x) \leq 0 \; j = 1,2, \ldots, n$

where x is the vector of design variables which represent the optimum layout problem, n is a set of inequality constraints of the form $g\,j(x) \leq 0$ $(j=1, 2, \ldots, n)$, and $xi \leq 0$ $(i=1,2, .., k)$, where k is a set of decision variables.

4.1.2 The objective function

The problem is to determine the optimum values of the variables:

- Windows (relation hollow - full in façade)
- Functional house corresponding dimensions and areas
- Building materials
- Thermal insulation (types, placing and thickness)

Covering of these variables can minimize the overall building cost, energy used, while satisfying the planning constraints, given the construction costs of the walls, windows, building materials, thermal insulation, house function, etc. The values of the variables provide the housing and cost and comfort. With the procedure described in the previous sections, the problem of generating the geometry has been solved. Thus, given a topology, the dimensions of a layout can be obtained which satisfies a number of constraints while minimizing the construction cost. Improved Move Limit method of Sequential Linear Programming provides a convenient and efficient method to solve dimensioning problems which are nonlinear programming problems (Rekha Bhowmik (4. april 2008)).

4.2 The main involvement factors

4.2.1 Enhancement of outdoor energy allocate by ameliorate of local microclimate

The first step towards a passive and low energy housing strategy is to create a competent and suitable local microclimate, which can be supported by handling the power of a negative climate variety (Georgi NJ, Zafiriadis (2006)). Existing winds, sun, noise and sources of pollution all can affect the environmental comfort level of user of open spaces and architectural spaces. Every residential site is a site definite as to its location, organisms, vegetation, solar access, and its microclimate (D. Pearlmuttere1993).

4.2.2 Interior energy allocate in the house by assign it such as cascade

We have to create a cybernetic system to calculate energy in the house to be an efficient and employ such cascade. Energy in the building must be allocate throughout regarding of thermal zones in the building by utilized the energy in diverse house functional spaces such as cascade.

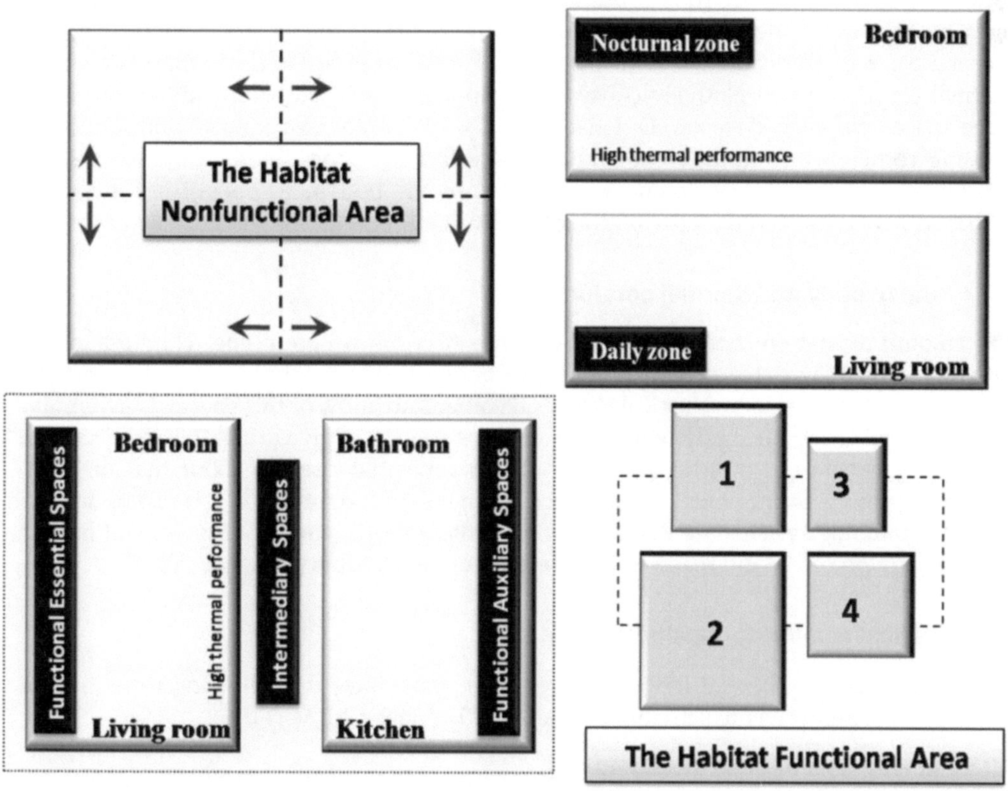

Fig. 3. The idea of house functional zoning

The vital step on passive and low energy housing is to reflect the energy distribution on the building form and volume, wherever the energy distribute be obliged to correspond the function and activity in those spaces.

4.2.3 Passive and low energy upon thermal house zone

There are three interior thermal zones;

4.2.3.1 Functional fundamental zone

This zone includes living space (bedrooms and living rooms). The optimal comfort temperature for these zones is between 22 -28 °C. The best place for functional essential spaces is in extremely center of the building.

4.2.3.2 Functional auxiliary zone

This zone includes kitchens and bathrooms. The optimal temperature for this zone is between 18- 28 °C. That means a 28°C for bathrooms and 18 °C for kitchen. This zone is modest warm and can locate in the periphery of the house plane for creates a natural ventilation. To be beside functional essential spaces for create of radiant heat corresponding building functional schema.

4.2.3.3 Intermediary zone

This zone includes storage rooms, buffer spaces, transit spaces, such as loggers, balconies, terraces, basements, etc. The optimal temperature for this zone is less than 10 °C. House thermal zones represent an enclosed space in which the air is free to flow around and whose thermal conditions are relatively consistent. Sometimes temperatures in different parts of large spaces can vary. (Watson. D. Labs, K. 1983). Well-organized passive and low energy housing recognizes these differences and creates thermal zones for the different building functional spaces. Thermal zoning tries to ensure the best match possible between the distribution of room and the distribution of the available energy.

4.2.4 Human body and thermal comfort

The amount of heat our bodies produce depends on what we are doing. The human body operates as an engine that produces heat. Our bodies turn only about one-fifth of the food energy we consume into mechanical work. The other four-fifths of this energy is given off as heat or stored as fat. The body requires continuous cooling to give off all this overload heat. When that person is sitting at a desk, the heat generated rises to about that of 100-W. Buildings provide environments where people can feel comfortable and safe. To understand the ways building systems are designed to meet these needs, we must first look at how the human body perceives and reacts to interior environments (Corky Binggeli 2003).

4.2.5 Thermal comfort for healthy habitat

- Under a healthy habitat indoor environment, must have the following recommended thermal comfort where activities are easy ≈ (70 W/m2 = 1.2 MET).

4.2.5.1 In winter conditions (heat required)

- Assuming a dress with a clo-value of 1 (0.155 m2 • K / W), obtained following conditions:
- Operational temperatures have to be between 20 - 24 ° C.
- The difference in the vertical air temperature between 0.1 m - 1.1 m above the floor (ankle and head-height) should be less than 3 ° C.

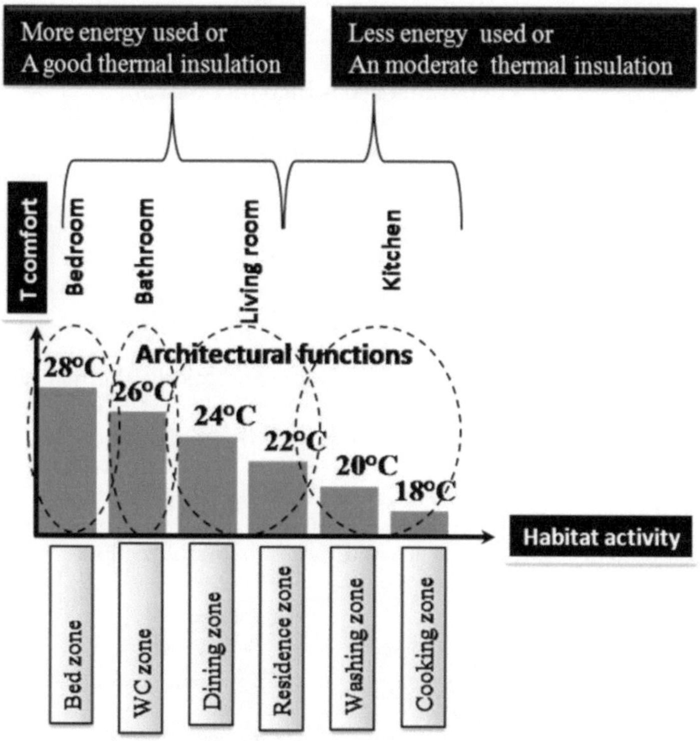

Fig. 4. The energy hierarchy allocates in a habitat functions

- The floor surface temperature should be between 19 - 26 ° C (floor heating systems can be sized for 29 ° C).
- Indoor Ambient air velocity should be less than 0.15 m / s.
- Radiation temperature by asymmetry form in which come from windows and other cold vertical surfaces should be less than 10 ° C (relative to a small vertical plane 0.6 m above floor).
- Radiation temperature asymmetry due to a hot (heated) ceiling should be less than 5 °C (relative to a small horizontal plane 0.6 m above the floor).

4.2.5.2 In summer conditions (cool required)

Assuming a dress with a clo value of 0.5 (0.078 m2 • K / W), obtained following conditions:

- Operational temperatures have to be between 23 - 26 ° C.
- The difference in the vertical air temperature between 0.1 m - 1.1 m above the floor (ankle and head-height) should be less than 3 ° C.
- Ambient medium speed should be less than 0.25 m / s.

The building envelope is the transition between the outdoors and the inside, consisting of the windows, doors, floors, walls, and roof of the building. The envelope encloses and shelters space. It furnishes a barrier to rain and protects from sun, wind, and harsh temperatures. Entries are the transition zone between the building's interior and the outside world. (Corky Binggeli .2003).

Habitat function	Thermal care level
Living room	80% of the area needs high thermal care
	20% of the area needs middle thermal care
Bed room	85% of the area needs high thermal care
	15% of the area needs middle thermal care
Kitchen	20% of the area needs high thermal care
	80% of the area needs low and middle thermal care
Bath room	80% of the area needs high thermal care
	20% of the area needs middle thermal care

Table 1. The requirement of a thermal care in different habitat functions

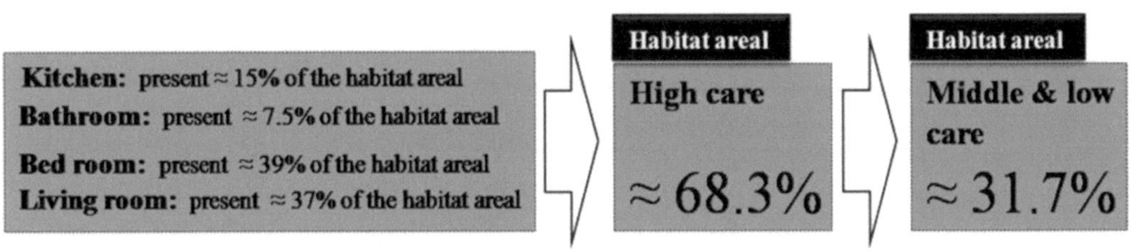

Fig. 5. The resulting of an optimal habitat thermal care requirement

68.3% of a habitat area needs a *"high thermal care"* to get an optimal human functions performance.

4.2.6 Energy and human metabolism and activity

The energy is used for growth, regeneration, and operation of the body's organs, such as muscle contraction, blood circulation, and breathing. It enables us to carry out our normal bodily functions and to perform work upon objects around us. The normal internal body temperature is around 37°C. (Ashley F. Emery 1986) The internal temperature of the human body can't vary by more than a few degrees without causing physical distress. The architect and engineer can establish the propose conditions by listing the variety of acceptable air and surface temperatures, air motions, relative humidifies, lighting levels, and background noise levels for each activity to take place in the housing. A schedule of operations for each activity is also developed.

4.2.7 Clothing and thermal comfort

In the greater part of cases, building inhabitants are inactive or slightly active and be dressed in classic indoor clothing. Clothing, through its insulation properties, is a vital modifier of body heat loss and comfort. See fig 6

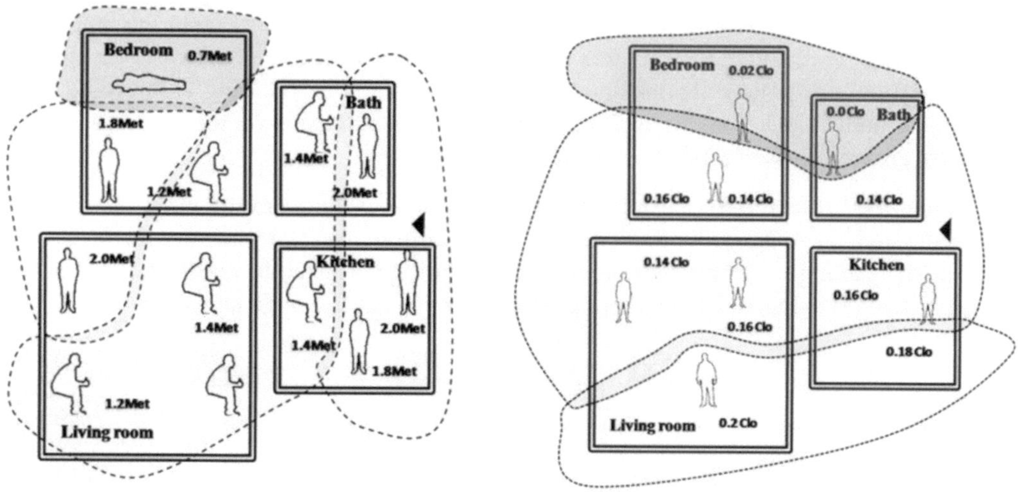

Fig. 6. House typical activity and clothing and functional zoning

The insulation belongings of clothing are a cause of the small air pockets alienated from each other to prevent air from migrating through the material (Freeman III; A Myrick 1993). In the same way, the well, soft down of ducks is a poor conductor and traps air in small, restricted spaces. In general, all clothing makes employ of this standard of trapped air within the layers of cloth fabric. Clothing insulation can be explained in terms of its *clo* value. The clo value is a numerical symbol of a clothing ensemble's thermal resistance. (1 clo = 0.155 m^2_°C/W).

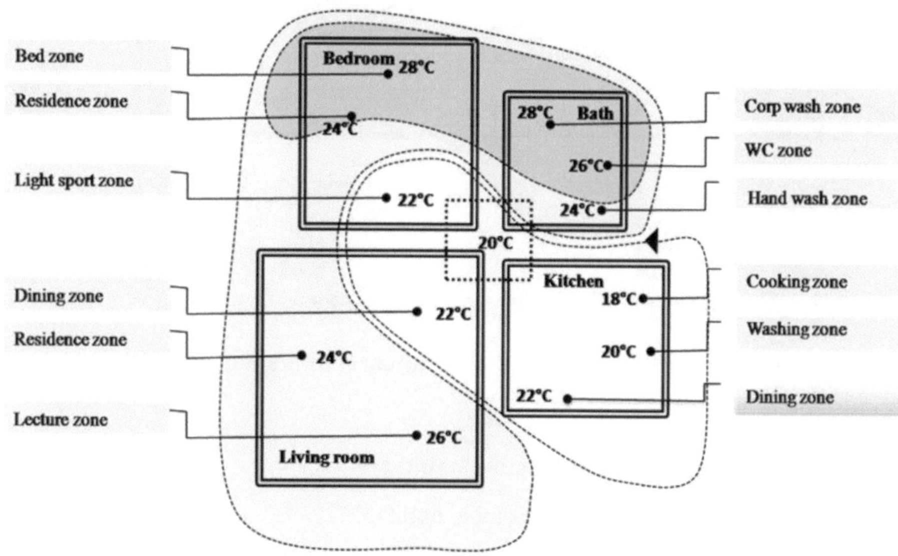

Fig. 7. Thermal analysis and house functions allocate

4.3 Human activity, clothing, human comfort, and architectural program

Each of us has our own preferred temperature that we consider comfortable. Most people's comfort zone tends to be narrow, ranging from 18°C to 24°C during the winter. Our body's internal heating system slows down when we are less active, and we expect the building's heating system to make up the difference. The design of the heating system and the quality of the heating equipment are major elements in keeping the building comfortable. Air movement and drafts, the thermal properties of the surfaces we touch, and relative humidity also affects our comfort. (Corky Binggeli .2003).

The human body has three mechanisms to preserve this fine temperature range. The first is heat generated inside the body, the second is by acquisition heat from surroundings, and the third is by gaining or losing heat to the surroundings. The body automatically makes constant changes to manage these three mechanisms and control body temperature.

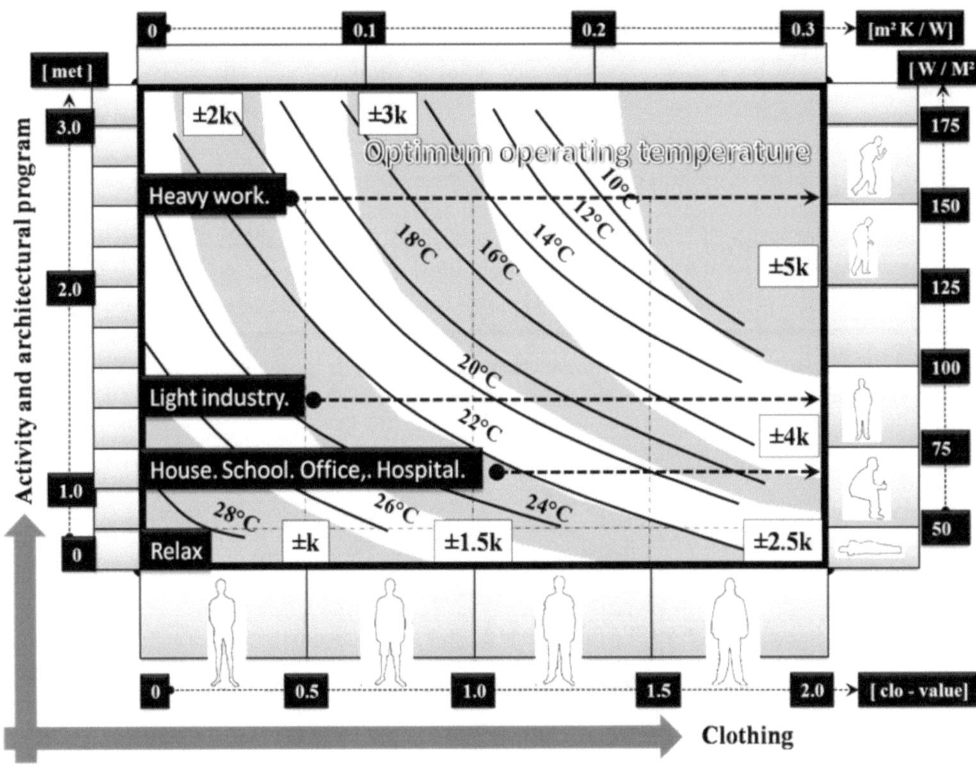

Fig. 8. The interaction between energy, activity, human comfort and architectural programs (Archcrea instate)

4.3.1 Surfaces temperature in comfortable habitat

For determination the Interior T in comfortable habitat

$$1. \text{MRT} = \frac{\sum T + \theta}{360} = \frac{T1\theta1 + T2\theta2 + T3\theta3 + +Tn\theta n}{360}$$

Where;

T = surface temperature
θ = surface exposure angle (relative to occupant) in degrees.

Actor Position	House Functions			
	Living room	Kitchen	Bath room	Bedroom
A	23°C	19.33°C	25°C	23.3°C
B	23.6°C	20°C		24°C
C	23.58°C			

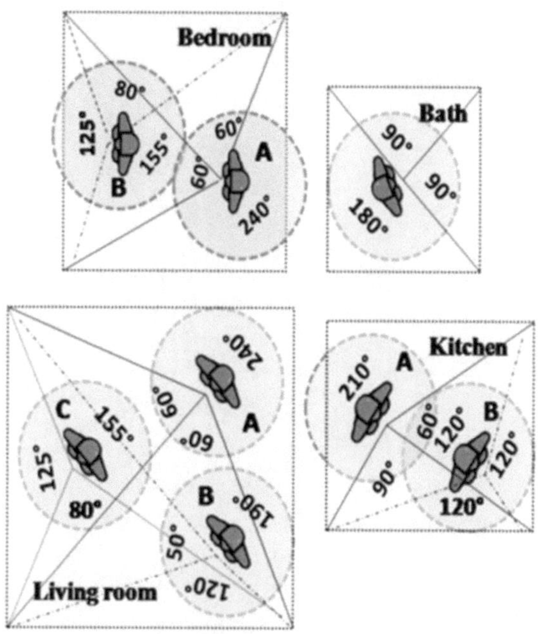

Table 2. The surface temperature for different actor position in thermal comfort spaces

The trends of how people spend their own time change from year to year. However, it contains broadly the same ingredients: a chance to escape from the city, to be alone or to be with other people, to be close to nature, and to relax and enjoy oneself (Jensen, C.R. and Guthrie, S.P. 2006). The oxford dictionary of science define adaptation such as "any change in the structure or functioning of an organism that makes it better suited to its environment". It is the evolutionary process whereby a population becomes better suited to its habitat (Bowler P.J. 2003). A human is adapted to the surroundings of the habitats in which he live. This process takes position over several generations, (Patterson C. 1999) and is one of the vital phenomena of biology.

This may be defined as a variable system of functional (structural) complexes and coordinates. The human body has three mechanisms to preserve this fine temperature range. *The first* is heat generated inside the body, *the second* is by acquisition heat from surroundings, and *the third* is by gaining or losing heat to the surroundings.

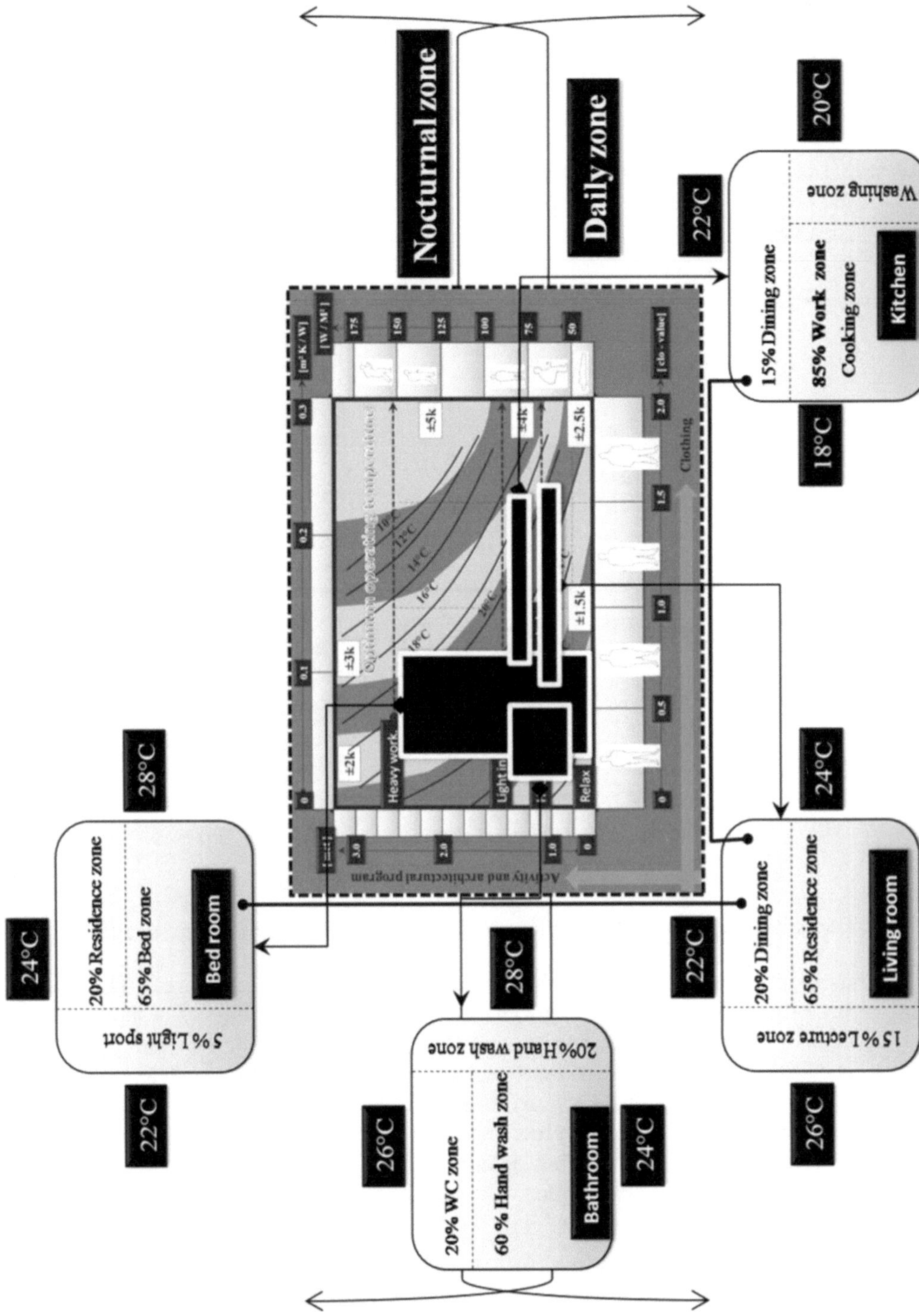

Fig. 9. Energy distribution in different habitat functions (Archcrea institute)

The body automatically makes constant changes to manage these three mechanisms and control body temperature.

4.3.2 Passive bioclimatic housing model

4.3.2.1 Windows

a. Passive window

Window plays a vital role in passive buildings classification in two ways:

- It diminish heat loss in spite of their a huge areas of glazing
- It permits the sunlight to create extra heat through the glass

We turn out to be aware of a modest loss of light-transmission and a slight brown tinting of the light due to the second layer of LE coating. (Craig A. Langston, Grace K. C 2001). Today, we can see the new models of energy low windows with 0.7 W/m^2 °C.

a.1. Energy manipulate on window

That represent regulates of the input and output energy, where the window's layers must be sufficient to limit the heat transfer in a dynamic system or limit the temperature change, with time, in a static system. The uncomfortable change of energy in temperate climate is in the winter in direction inside outside. In general, heat always flows from warmer to cooler. This flow does not stop until the temperature in the two surfaces is equal. Heat is "*transferred*" by four different means: conduction, convection, radiation and infiltration. Insulation decreases the transference of heat. Well designed and protected windows improve comfort year round and reduce the need for heating in winter and cooling in summer. In reality, the serious lighting designer cannot take any notice of the energy implications of window choices. New technologies help to resolve the historic problem of the transaction between windows that reflect unwanted solar gains in the summer and those that admit a maximum quantity of useful light. Well designed windows and shading devices allow solar heat gain in winter and shade and ventilation in summer while providing enough day lighting. Solar gain achieved by heaving 60% of the building's windows orientated correctly can reduce the heating load of a building by ≈22%. Shutters can be used to control the amount of heat (and light) transferred through the glass, and box pelmets and long wide curtains can limit air movement over the glass and prevent draughts (Assad Z. K. Almssad 2005).

a.2. Thermal window's functions

The surface temperature of single-glazing, for example, will be extremely bad insulating to external temperatures. That is to say interior surface temperature of double glazing will be much warmer but still significantly lower than interior temperature. Frames, which can take an area 10-30% of a typical window, also have perceptible effects; surface temperatures of insulating frames will be much warmer than those of highly conductive frames. Warmer glass surface temperatures translate into more comfortable spaces or occupants during the winter because comfort is a function of radiant heat transfer among people and their surroundings.

For optimal thermal treatment of windows we must know the components of windows that play a thermal role correlated to energy exchange interior exterior in cold winter and contrary in summer. These are pane glass and frames. By perceptive how windows control thermal comfort, windows designer can create an optimal resolution of windows, which collaborated welcoming with environment. Windows in residential building consume approximately 2% of all the energy used in industrials countries. Well organized windows can greatly improve the thermal comfort on houses during both heating and cooling seasons. Therefore Windows play significant role in the design strongly affect their energy use. Condensation on the windows may be a sign of heat loss. A damp area around the window from the exterior is another sign of heat loss. During the winter, a typical window loses up to 10 times more heat than an equivalent area of an outside wall or roof. Windows can account for up to 30 % of the heat loss from a conventional building, adding significantly to heat cost. Drafts, window condensation and mould can also affect our comfort and indoor air quality. Sustainability is a wise approach to the way we live. And using energy in a more sustainable way is a part of this approach (Assad Z. K. Almssad 2005).

b. Bioclimatic windows(sizing, orientation and habitat functions)

For getting an optimal lighting, which is very important for passive and low energy housing, the house functional orientation must matches precisely the activity level on this spaces which keep up a correspondence with the better lighting for specifically function. The lengthiest window dimension is for activity in which needs more precision, such as study and living area

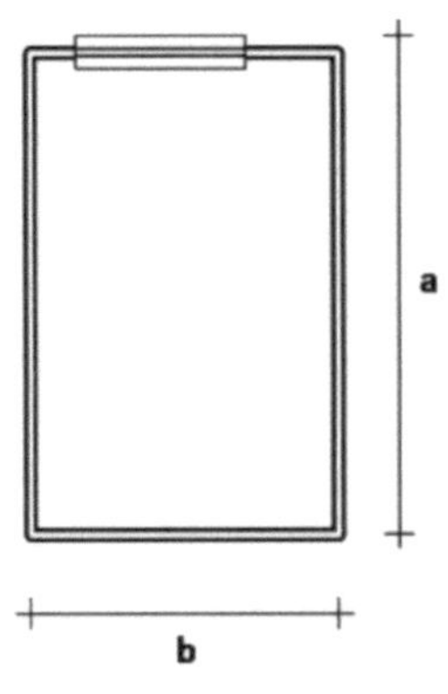

Fig. 10. Functional space dimensions

c. Improvement of windows thermal and optical functions

Heat loss in winter is a severe problem in which architects, engineers, must get basically explanation, for the thermal part of windows problems, in the following part of this research we will try to describe these problems and suggest the best solution.

- Air such as intermediary thermal layer

Energetic role of intermediary thermal layer is between cold exterior spaces and comfortable interior living spaces, therefore the essential role of intermediary layer is such a thermal buffering. For the parent building, a thermal intermediary layer represents a reduction in heat loss/gain through the windows. This is because these elements lose/gain heat to a space which is at a higher or lower temperature than the outdoors. The magnitude of the

The house functions		The best orientation		Relation Length (a) - width (b)	Minimum window length (meter)	Maximum window length(meter)
main function	sub function	Specifications	Overall			
Living room	Residence zone	East, south, south west	South, South-west South-vest East	a < b	0.11 (a x b)	0.17 (a x b)
	Dining zone	South and South west		a = b	0. 12 (a x b)	0.18 (a x b)
	Lecture zone	East, south, south East		a > b	0.13 (a x b)	0,19 (a x b)
Bedroom	Bed zone	East, South - east	South South-east East	a < b	0.080 (a x b)	0.105 (a x b)
	Residence zone	South-east, east, south		a = b	0.085 (a x b)	010 (a x b)
	Light sport	South, south - east, east		a > b	0,90 (a x b)	0.12 (a x b)
Kitchen	Cooking zone	North, North - east	North North-east	a < b	0.070 (a x b)	0.95 (a x b)
	Dining zone	South west, North		a = b	0, 075(a x b)	0.10 (a x b)
	Washing zone	North, north-west		a > b	0.078 (a x b)	0.11 (a x b)
Special: bathroom	Wishing hands and body	North, North west		a , b	0.36 m²	0.72 m²
Special: Service zone, and other auxiliary functions	Circulation, stores, recreations	North, South, West, South-West, North - west	North for temperate climate and South for hot climate	a , b	0.01 m²	0.36 m²

Table 3. The optimal windows orientation, length in bioclimatic house (Archcrea institute 2010)

heat loss/gain reduction depends on the temperature of the intermediary space. The optimal temperature for this layer is around 8°C for buildings in winter season. The air in intermediary thermal layer must have a higher temperature than exterior on winter and reverse on summer.

- Heating recovery system

This system consists of two separate air management systems, one collects and exhausts stale indoor air, and the other draws in fresh outdoor air and distributes it right through the house. That means using of stale indoor air eliminate in air refreshing process and charging the outdoor fresh air set up with optimistic energy these process can be present by using mechanical ventilation. The incoming air can also heat or cold the internals spaces of a building in passive buildings, thus resulting in technically simple solutions for cool/heat provide system.

- Window's Light shelves

The light shelf itself is a simple device that is installed inside the window. In most applications, it must be combined with other devices to avoid glare from sunlight incoming the lower portion of the window. The glare problem is avoided by this system that limits the use of diffusers to make use of day lighting through windows. It also provides the unique advantage of variable the light from the window so that it comes from a more overhead direction, humanizing the quality of illumination. For an efficient function of light shelf system, it requires direct sunlight. The windows should face towards the sun for a large portion of the time that the space is occupied. Tinted or reflective glazing may very much reduce the potential benefit of light shelves, or make them uneconomical. These types of glazing typically block about 70-80% of incoming sunlight. In some cases the system may be used with glazing at lower heights where people cannot get close to the glazing. As with

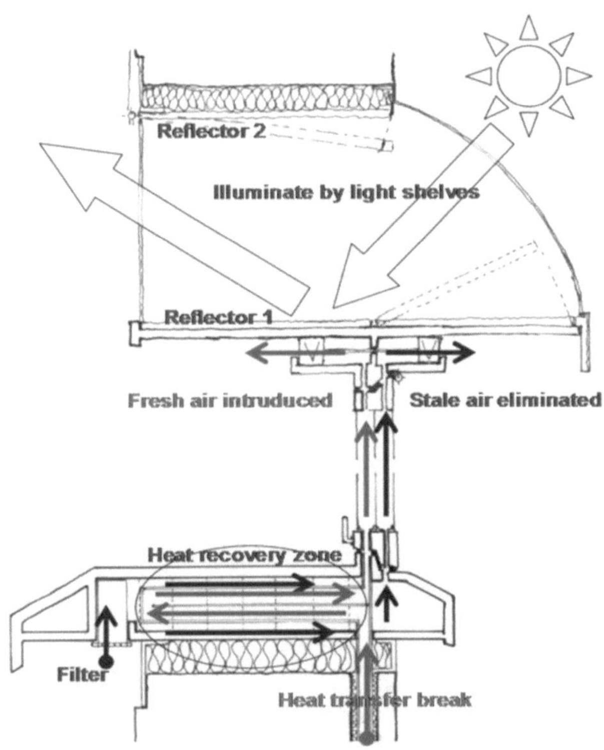

Fig. 11. Optimal window function

any kind of day lighting, the electric lighting must be arranged and controlled so that it can be turned off to exploit the daylight provided by the light shelf system. The location is apparent for tall windows, where the light shelves can provide deeper penetration than day lighting that achieved by shading windows. This is because light shelves can throw all the energy of direct sunlight into the space. In contrast, using shading to tame sunlight for day lighting leaves most of the potential day lighting energy outside the building. For a well-organized function of light shelf system that needs;

- The good treatment of windows. A window must be exposed to direct sunlight to be an applicant for a light shelf. Effective day lighting by any method is still infrequency. It is impossible to communicate the visual effect of day lighting by words or figures.
- The simplest materials and function of light shelf such reflector. It could be as simple as aluminum foil taped to a piece of cardboard.
- The distribution function of day lighting is from the portion of the window that extends above the light shelf. The window must face towards the sun for a large part of the time, and it cannot be shaded by outside objects. If the window glazing is tinted or reflective, the day lighting potential is reduced substantially.
- The ceiling is another manner of light distributor, where it can receive from light shelf. The height and orientation of the ceiling and the diffusion characteristics of the ceiling distributes the daylight.

4.3.2.2 Insulations upon house climatic screens

a. External doors

Insulating or replacing external doors can assist to diminish draughts and heat loss at building. The typical door for a low energy building has a U-value of 1.0 W/m^2 C°. This can be improved by adding an extra internal door swinging into the room. Otherwise doors with U-value of 0.8 W/m^2 C° must be sourced (ECN).

b. Walls

A wall is a frequently solid structure that describes and sometimes protects an area.

c. Roofs

The thick roof takes away too much headroom except in the widest building type with increased roof pitch. The passive upgrade consists of special cold-bridge free roof truss without vertical enforcements. U-value 0.082 W/m^2 C° (Amjad Al-musaed 2004)

4.3.3 Other thermal assignment

4.3.3.1 Thermal bridging effects

Thermal bridges are divisions through the material of significantly inferior thermal resistance than the rest of the building. These happen chiefly in the region of openings and at joint of walls- floors and walls roofs. Building element intersections: The linear thermal transmittance to exterior has to be below 0.01 W/m K°. (Sugawara, M. and A. Hoyano. 1996).

4.3.3.2 Air tightness

Tight buildings reduce energy costs by keeping in the comforted air conditioned air. But tight buildings without adequate ventilation catch humidity and pollutants so they feel unventilated, aggravate allergies and source general discomfort for house occupants. Moisture damage to windows and other parts of the house covering can result when humidity is excessively high. New houses, additions and even remodeling projects are far more airtight than they used to be. Building a tight house to today's standers can engrave the overall heat gain/loss by 25-50%. This is progress; a tight house is more comfortable, because it is less drafty and less expensive to cool/heat, because the energy man pay for stays in the house longer. on the other hand, a tightly constructed house needs also a mechanical ventilation to keep the air inside fresh and stop the buildup of indoor air pollutants such as excess moisture, carbon dioxide, formaldehyde and various volatile organic components found in buildings materials, paints, furnishing, cleaning products and smoke.

Air tightness is a concept of control of the ventilation airflow rates. It creates achievable to minimize energy use while maintaining a high-quality indoor environment. All windows and doors are required to meet the required air-leakage values. < 0.6 air changes /h at n 50pa. The total energy performance of a passive building is, to a large degree, needy on how airtight the building is (Sugawara, M. and A. Hoyano. 1996).

4.3.3.3 Underground thermal inertia

Using of the underground constant temperature can be useful for architects and designer because the temperature is between 8 °C - 13 °C, and 3 m above the earth, can help us to find a controlled thermal flux from underground to building elements by means of a determinant tube canals. (Amjad Al-musaed 2004).

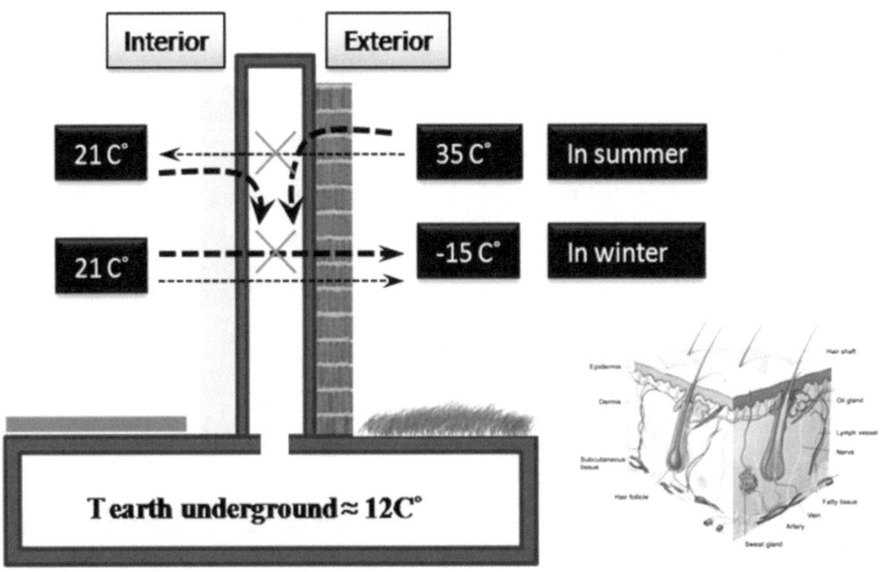

Fig. 12. The earth °C underground (using model)

4.3.4 Passive heating systems and strategies

The systems can be (direct- indirect). Passive heating strategies in particular make use of the building components to collect, store, and distribute solar heat gains to reduce the requirement for space heating. We have to create a variety of systems of solar heating systems and a clear description of a passive heating strategy.

Fig. 13. Passive heating process

Passive heating systems correspond to an environmentally friendly method of a human healthy building.

5. References

Amjad Almusaed 2010, Biophilic and bioclimatic Architecture, Analytical therapy for the next generation of passive sustainable architecture, Springer- Verlag London, UK

Almusaed 2004, Intelligent Sustainable Strategies upon Passive Bioclimatic Houses, Post Doctorial research, the architect school of architecture, Aarhus, Denmark

Ashley F. Emery 1986, http://courses.washington.edu/me333afe/Comfort_Health.pdf

Assad Z. K. Almssad 2005, Bio-sustainable ecological windows, the world sustainable building conference, Tokyo Japan

Bowler, P.J. (2003), Evolution: the history of an idea (3rd ed.), University of California Press p12

Corky Binggeli .2003, Building systems for interior designers, John Wiley & Sons, Inc. Pp 17-24

Craig A. Langston & Grace K. C. Ding 2001, Sustainable practices in the built environment, Plant tree, Second edition, GB Maulbetsch

Freeman III, A Myrick 1993. The measurement of environment and resources values: Theory and methods: Resources for the future, Washington DC, USA

Georgi NJ. Zafiriadis (2006), The impact of park trees on microclimate in urban areas. Urban Ecosystem

Jensen, C.R. and Guthrie, S.P. 2006, Outdoor Recreation in America (6th ed.), Human Kinetics: Champaign, Illinois.

Pearlmuttere 1993 Roof geometry as a determinant of thermal zone, Architectural science review, Vol. 36, No. 2

Rekha Bhowmik (4. april 2008) Building Design Optimization Using Sequential Linear Programming, Journal of Computers, Vol. 3, No. 4, April 2008.

Robert Hastings & Maria Wall, 2009, sustainable solar houses, strategies and solutions, Earthscan publisher, UK, and USA, p 63

Sugawara, M. and A. Hoyano. 1996. "Development of a Natural Ventilation System Using a Pitched Roof of Breathing Walls." Pp. 717-722 in Proceedings of the 7th International Conference on Indoor Air Quality and Climate - Indoor Air 96. Volume. 3

Watson. D. Labs, K. 1983, Climatic Design: Energy efficient building principles and practices. McGraw-Hill; New York

Part 2

Energy Efficiency upon Passive Building

3

Traditional Houses with Stone Walls in Temperate Climates: The Impact of Various Insulation Strategies

Francesca Stazi, Fabiola Angeletti and Costanzo di Perna
Polytechnic University of Marche
Italy

1. Introduction

The European Directive 2010/31/EU states that the measures to improve the energy performance of new and existing buildings should take into account climatic and local conditions as well as indoor climate environment giving the priority to the summer period. The Directive introduces the concept of thermal capacity that is very important in temperate climates during the summer phase.

The adoption of North-European super-insulated structures, characterised by low transmittance, involves the use of considerably thick thermal insulation both when designing new developments and when energy retrofitting existing traditional buildings.

The introduction of external insulation coatings has led to a change in the relationship between architecture and climate, resulting in an "international" envelope and the loss of local features; also, in temperate climates, because of indoor overheating it has led to various problems as regards summertime comfort.

The deeply rooted link between regional traditional architecture and specific climates has been underlined by many researchers (Givoni, 1998; Grosso 1998).

Many authors have carried out quantitative studies on the behaviour of traditional buildings in different seasons, showing the capacity of these buildings to secure indoor comfort by means of passive control systems. Experimental studies carried out on traditional houses in India (Dili et al., 2010) Korea (Youngryel et al., 2009), northern China (Wang & Liu, 2002), Japan (Ryozo, 2002), Nepal (Rijal & Yoshida, 2002), and Italy (Cardinale et al., 2010) are of particular interest.

Some authors have experimentally compared the behaviour of traditional and modern buildings, not only showing how the former are preferable both from the point of view of comfort and in terms of energy saving (Martin et al., 2010) but also identifying the limitations of energy saving legislation which tends to encourage the latter (Yilmaz, 2007; C. Di Perna et al., 2009).

Few authors have assessed the impact of energy saving strategies on traditional architecture (Fuller et al., 2009).

The present chapter focuses on the yearly behaviour of a traditional farmhouse in a temperate Italian climate and analyses the impact of alternative energy saving strategies on summer comfort and winter consumptions.

The aim of the current study was to:

- identify combinations of thermal insulation interventions which: optimise winter energy saving and summer internal comfort without modifying the close relationship between architecture and specific climate typical of traditional buildings; respect the building material consistency and the façades aesthetic appearance.
- compare the performance of this traditional architecture (after the retrofit intervention) with that of a modern building, of the type encouraged by new energy saving legislation (lightweight and super-insulated).

To that aim a series of monitoring activities in summer and in winter were carried out to investigate the internal environmental conditions and to calibrate a simulation model with the software Energyplus. This model was used to assess the impact of various energy-saving strategies on winter energy consumptions and summer comfort with the method of Percentage outside the comfort range (EN 15251:2007-08).

2. Research methods

This research study involved an experimental campaign to analyze the behaviour of a traditional building in different seasons and sunlight conditions as well as a series of simulations in order to extend the comparison to other types of walls representing either alternative energy retrofit solutions or newly built envelopes. The study was carried out in the following stages:

- Experimental campaign
- Dynamic simulation using Energyplus software
- Calibration of the simulation model by comparing the calculated and measured values
- Parametric analysis so as to assess different retrofit scenarios.

2.1 The case study

The building is located between the Apennines and the Adriatic Sea. The traditional rural architecture is characterized by passive control elements such as stone walls with high thermal inertia and small windows: these characteristic features represent the main elements of architectural quality. However, these buildings often have problems during the summertime due to low temperatures for considerable heat loss to the ground.

This study focused on a traditional Italian farmhouse (fig.1) in the city of Fabriano situated 325 m above sea level and with 2198 degree days (longitude 13°37′; latitude 43°62′). The farmhouse has two floors above ground level and is characterized by a surface area to volume ratio of 0.60; it is oriented with its longitudinal axis rotated 45° clockwise with respect to the north-south direction. The building is currently used as a private residence with one apartment on the ground floor and two apartments on the first floor. The load-bearing structure is built in plastered masonry consisting of local stone and crushed bricks bound together by sand and lime mortar. These walls are 60 cm thick at ground floor level

Fig. 1. External view of the building

and 40 cm thick on the first floor. Both the walls have high thermal inertia. The cement floor slab is about 15 cm thick and lies directly on the ground, while the recently replaced intermediate floor is made up with brick and concrete. The roof is made of wood and is covered with tiles.

2.2 Experimental study

Monitoring of the building was carried out both during the summertime (23-30 July 2008) and in the winter (8-14 January 2009), in order to analyze the performance of the external stonework in two different weather conditions. The monitoring involved (fig. 2 and table 1):

1. Analysis of the outdoor environmental conditions by means of a weather station which measured the speed and direction of the wind, the solar radiation, the temperature and the relative humidity.
2. Analysis of the indoor environmental conditions by means of two data systems (one located in the southeast side and one in the north-west side) which recorded: the mean radiant temperature, the temperature and the relative humidity of the air.
3. Analysis of the thermal performance of the vertical cladding using thermo-resistances to measure the indoor and outdoor surface temperatures of the masonry.
4. Analysis of the heat transfer through the walls using a flow-meter.

During the summertime monitoring period the householders agreed to regulate the ventilation keeping the windows open in the morning (8:00- 10:00) and in the evening (18:00- 23:00). Some rooms were monitored with the windows always closed so as to determine the impact of natural ventilation on indoor comfort.

During the winter monitoring period heating was turned on from 7:00 to 9:00 and from 13:00 to 14:00 and from 19:00 to 23:00 (temperature set-point 20°C). Some unused rooms on the north side of the building were never heated.

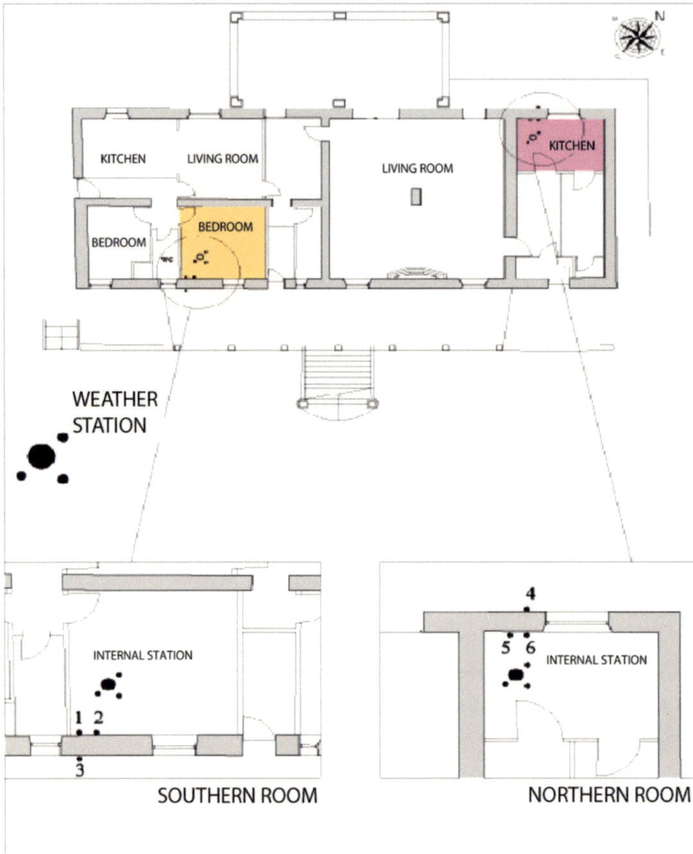

Fig. 2. Plan of the building with the instruments' position

SUMMER	23 July	24 July	25 July	26 July	27 July	28 July	29 July	30 July
MAX Temp	25,2 °C	28,8 °C	32,2 °C	32,7 °C	29,0 °C	31,2 °C	31,6 °C	27,1 °C
MAX REL. HUMIDITY	45%	46%	46%	46%	46%	46%	46%	45%
MIN Temp	19,1 °C	19,2 °C	21,8 °C	24,1 °C	23,5 °C	22,5 °C	21,0 °C	22,3 °C
MIN REL. HUMIDITY	44%	39%	38%	38%	39%	39%	38%	39%
DAY	SUNNY	SUNNY	SUNNY	SUNNY	SUNNY	CLOUDY	SUNNY	SUNNY
MAX WIND SPEED	1,5 m/s	3,1 m/s	2,7 m/s	2,1 m/s	2,8 m/s	1,8 m/s	2,0 m/s	0,8 m/s
WINTER	8 January	9 January	10 January	11 January	12 January	13 January	14 January	
MAX Temp	4,7°C	4,8 °C	5,1 °C	10,1°	9,2°C	4,5°C	6,7°C	
MAX REL. HUMIDITY	100%	100%	89%	75%	65%	100%	100%	
MIN Temp	0,8 °C	1,9 °C	1,0 °C	-1,1°C	-2,2°C	3,0 °C	4,1 °C	
MIN REL. HUMIDITY	93%	88%	55%	37%	23%	56%	92%	
DAY	RAINY	SNOWY	CLOUDY	SUNNY	CLOUDY	FOGGY	FOGGY	
MAX WIND SPEED	2,1 m/s	3,3 m/s	3,2 m/s	3,5 m/s	3,1 m/s	4,8 m/s	6,0 m/s	

Table 1. Weather scheme of the monitored weeks

Uncertainty is evaluated as a percentage of accuracy and a percentage of resolution declared by the manufacturers. For the thermometric probes the acquired data had a tolerance of ±0.15 °C distributed equally for all the values measured, while in the case of the flow-meter the accuracy was determined by the formula 5%VL + 1 W/m^{-2} (VL = read value) and results in a mean square error which fluctuates over time.

2.3 Analytic study

Simulation of the case study in the real "as-built" state was performed using dynamic software (Energy Plus). Calibration of the virtual model developed in the dynamic state was carried out by comparing the indoor and outdoor surface temperatures and the air temperature recorded during the monitoring. This calibration involved the preparation of a weather file using the real weather condition data. It was subsequently necessary to take steps to modify the input values of the software (above all regarding the infiltration model) so as to reflect the real occupancy, ventilation and heating.

Parametric analyses were carried out on this corrected model, in order to analyse the relative impact of different types of energy saving solutions on summer comfort and winter consumptions. The retrofit interventions were identified according to the Italian standard (legislative decree 192/05), implementing Directive 2002/91/EC on the energy performance of buildings, and subsequent corrective decrees: for the climatic zone of interest, the regulation set a limit of 0,34 W/m^2K for the vertical opaque envelope, a value of 0,30 W/m^2K for the roof and a value of 0,33 W/m^2K for the ground floor. The following scenarios were compared (tab.2):

- Scheme A regards the "as built" state
- Scheme B regards the visible interventions, i.e. insulation of the vertical walls with an external 10cm thick low-transmittance "coating" of high density rock wool
- Scheme C regards the interventions that are not visible such as the thermal insulation of the intermediate and ground floor, with panels of high density rock wool: 3 cm thick for the intermediate floor and 4 cm thick for the ground floor; the thermal insulation of the roof (with panels of high density rock wool 11 cm thick); the replacement of single panes with low-e double glazing units (4/12/4), keeping the same window frames and wooden shutters;
- Scheme D regards the combination of the previous interventions (scheme B and C)
- Scheme E regards a newly built super-insulated lightweight building realized according to Italian legislation on energy saving (wooden structure and external single layer envelope of insulation material).

The energy consumption, the surface temperatures and the HOURS OF DISCOMFORT, calculated using the Method of Percentage outside the comfort range (Method A- Annex F- ISO EN 15251) were compared for the different simulated solutions.

3. Experimental results

3.1 Summertime monitoring

A comparison between the south-facing and north-facing walls (Fig.3) shows that the external surface temperature in the former reaches a maximum value at about 2 pm, while

	SCHEME A	SCHEME B	SCHEME C	SCHEME D	SCHEME E
Intervention					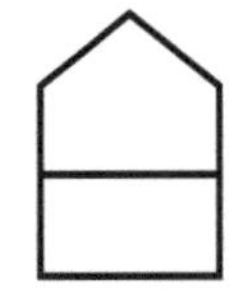
Vertical walls	**P1= as built stone wall**	**P2= external insulation**	**P1= as built stone wall**	**P2= external insulation**	**P3=new wall**
Thermal transmittance (W/m^{2}K)	1,92	0,33	1,92	0,33	0,13
Periodic thermal transmittance (W/m^{2}K)	0,14	0,01	0,14	0,01	0,10
Internal areal heat capacity k1 (kJ/m^{2}K)	206	201	206	201	27
Ground floor slab	**S1= as built floors**	**S1= as built floors**	**S2= insulation of the floors**	**S2= insulation of the floors**	**S3= new floors**
Thermal transmittance (W/m^{2}K)	2,63	2,63	0,32	0,32	0,32
Periodic thermal transmittance (W/m^{2}K)	0,86	0,86	0,01	0,01	0,01
Internal areal heat capacity k1 (kJ/m^{2}K)	226	226	240	240	240
Roof slab	**T1= as built roof**	**T1= as built roof**	**T2= insulation of the roof**	**T2= insulation of the roof**	**T3= new roof**
Thermal transmittance (W/m^{2}K)	4,49	4,49	0,29*	0,29	0,26
Periodic thermal transmittance (W/m^{2}K)	4,39	4,39	0,18*	0,18	0,17
Internal areal heat capacity k1 (kJ/m^{2}K)	30	30	88*	88	89
Glazing	**V1= as built single glazing**	**V1= as built single glazing**	**V2= double glazing**	**V2= double glazing**	**V3=double glazing**
Thermal transmittance (W/m^{2}K)	5,5	5,5	2,27	2,27	2,27

*in the scheme C1 the insulation of the roof regards the introduction of an insulated horizontal attic slab with thermal transmittance U=0,30 W/m^{2}K; periodic thermal transmittance Yie=0,30 W/m^{2}K; internal areal heat capacity k_1=13 kJ/m^{2}K. In that scheme the superior pitched roof slab has a thermal transmittance U=0,80 W/m^{2}K

Table 2. The intervention analysed.

the latter always presents a maximum value of surface temperature in the evening at about 6 pm, exceeding the values recorded in the south side. This behaviour is because the north-west side is struck by sunlight with a low angle in the final hours of the day.

The thermal peaks that occur on the external side of the walls are not present on the internal side and the internal surface temperatures are low, leading to discomfort.

On the sunny day the maximum range in temperature between the internal and the external façade is about 6°C to the south and 10°C to the north; the indoor surface temperatures throughout the day are around 20°C to the north and 22°C to the south.

On the cloudy day (28th July) the maximum temperature difference between the indoor and outdoor surfaces is 3°C to the south and 5°C to the north and the indoor temperatures throughout the day are around 21°C to the north and 23°C to the south.

The inner surface temperature in the north-facing room is about 2 °C colder than the south-facing one, both on the cloudy and on the sunny day.

Therefore, the difference in temperature found for the south- and the north-facing walls is constant (about 2°C) in all types of weather conditions: this depends on the fact that the thick exterior wall strongly delay the heat transfer from the external to the internal side; also internal surface temperature's trend is influenced by the internal air temperature (Fig. 4) and by the contribution of internal heat gains due, for example, to the solar radiation entering through the windows, the presence of people and equipments. The north-facing part of the building is not inhabited thereby eliminating all the internal heat gains.

The graph on Fig.3 shows a progressive weekly increase in internal surface temperatures. This depends on the high thermal inertia of the walls that involves a high capacitive resistance and therefore a significant delay (around a week).

The internal surface temperatures (fig.5) were compared in two south-facing rooms, in such a way as to have cross ventilation throughout the day in one case and the windows closed in the other.

Opening the windows during the night leads to a slight increase in the indoor temperatures at night since the outdoor air is warmer. During the daytime, natural ventilation leads to a reduction of about 0.5°C in the internal surface temperatures, except between 15.00 and 17.00 when the outdoor temperature rises. Throughout the daytime, in both cases, whether the windows are closed or open, the recorded value is less than 23°C bringing about an operative temperature always lower than the optimal summertime temperature.

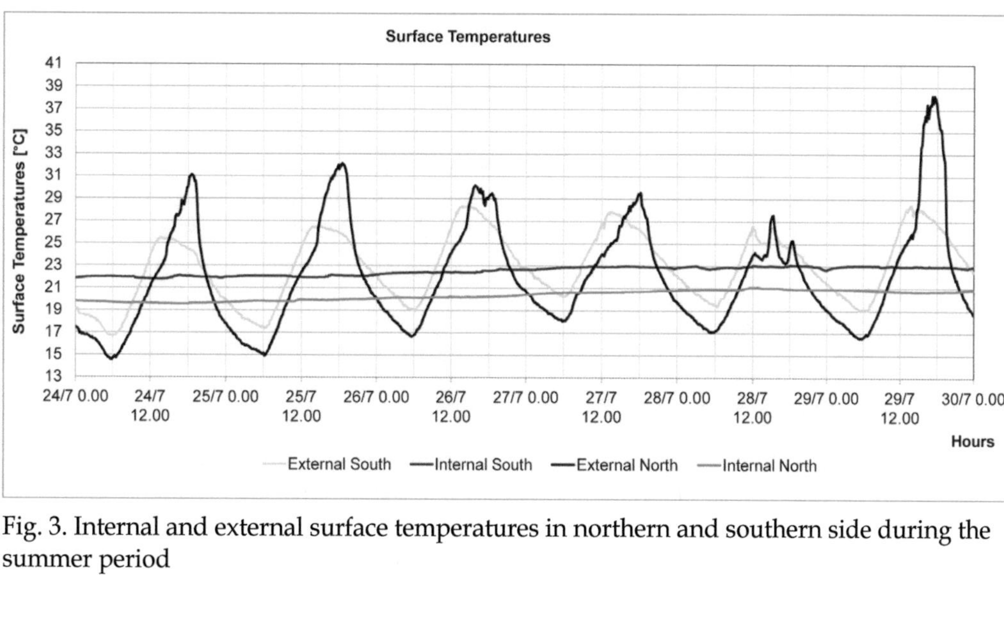

Fig. 3. Internal and external surface temperatures in northern and southern side during the summer period

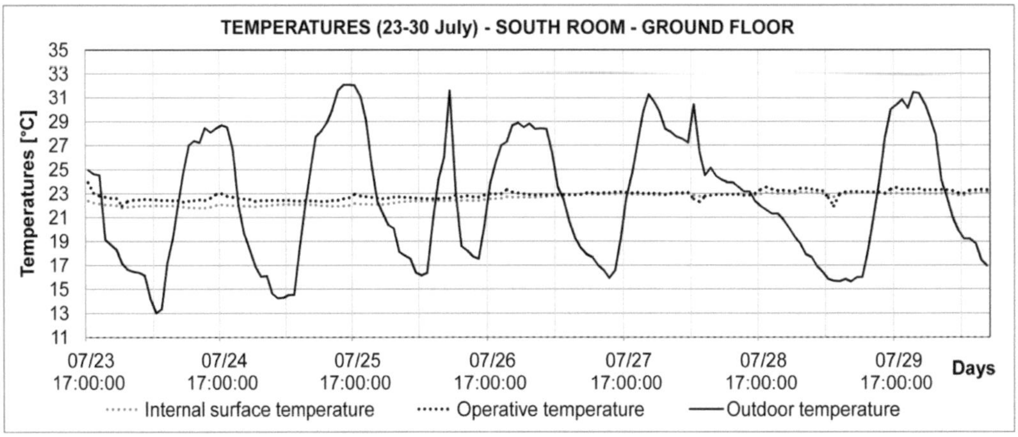

Fig. 4. Operative temperature on the southern room

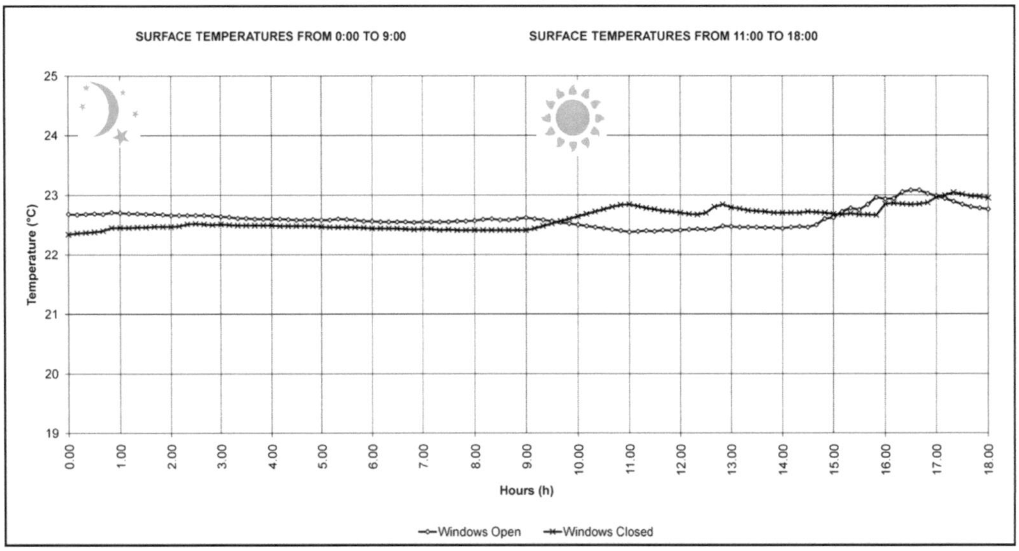

Fig. 5. Impact of ventilation on thermal inertia during the summertime. The graph shows the internal surface temperatures recorded in two rooms: one with windows opened and one with windows closed

3.2 Wintertime monitoring

The indoor and outdoor surface temperatures were compared for a north-facing and a south-facing wall (fig.6).

An analysis of the indoor surface temperatures shows that the south-facing room is about 15°C warmer than the north-facing one. This is due to the fact that the north-facing part of the building is not heated. In both cases (with heating on or off) the indoor surface temperatures have a very constant trend, without fluctuations.

As regards the temperatures of the south-facing wall, it can be observed that on both cloudy and sunny days, even with very different external surface temperatures, the indoor surface temperature is around 16° - 17°C: this value is very low and causes indoor discomfort.

The temperature variations in the south room are due to the influence of heating plants as it can be seen from operative temperature. (fig.7)

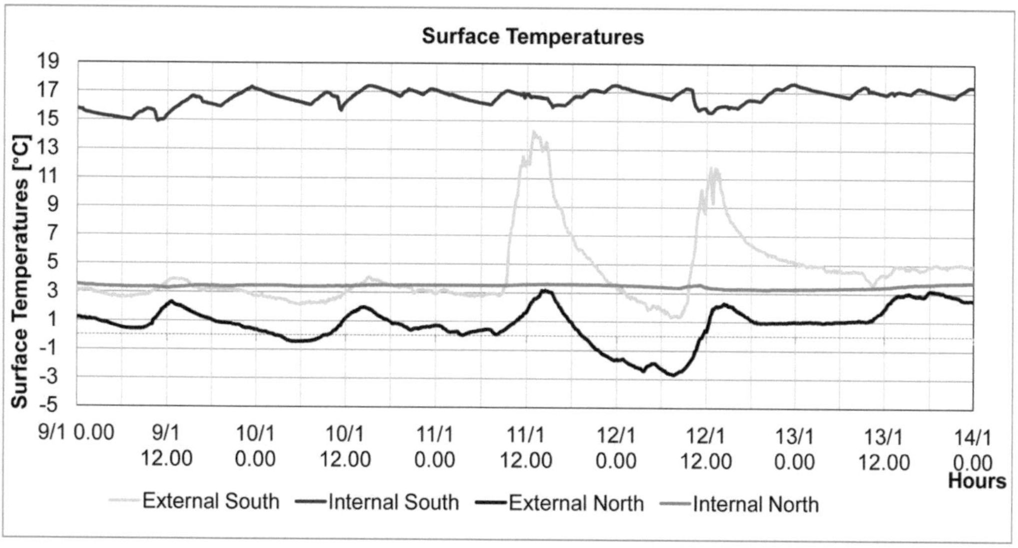

Fig. 6. Internal and external surface temperatures in northern and southern side during the winter period

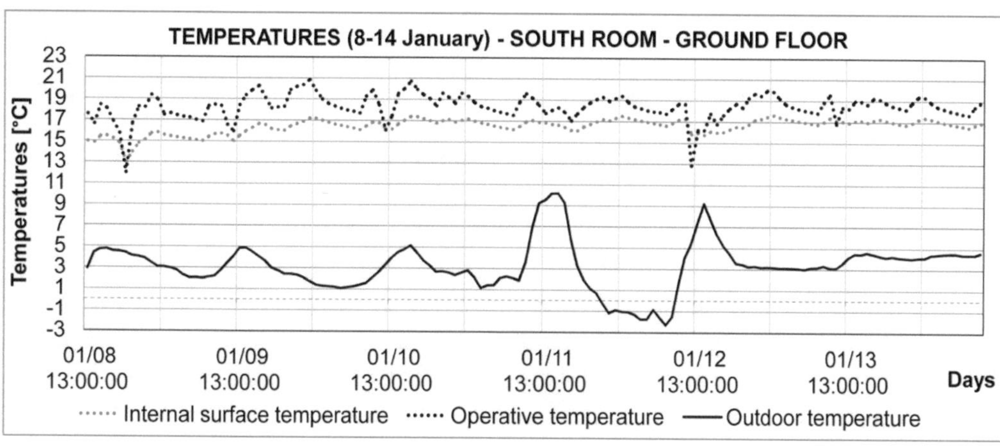

Fig. 7. Operative temperature in wintertime.

4. Analytic results

To analyze the relative impact of different types of energy saving solutions on summer comfort several simulation models were developed and compared.

We separated the interventions that do not affect the architectural image and the interventions that alter the architectural appearance of the building.

In the second stage the interventions on visible elements and interventions on the non-visible elements were combined and the building optimised according to the new regulations on energy saving was compared with a new lightweight super-insulated building.

The study of discomfort hours using the Method of Percentage outside the comfort range (Method A- Annex F- ISO EN 15251) gave the following results (Tab.3):

Interventions		Scheme A	Scheme B	Scheme C	Scheme C1	Scheme D	Scheme E
							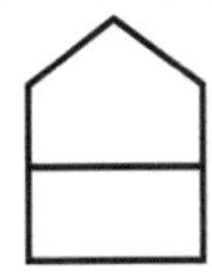
		As built state	*External coating*	*Insulation of floors and glazing*	*Scheme C+ attic floor insulation*	*Insulation of all constr. elements*	*New lightweight building*
Ground floor	overheating	0	0	0	0	0	0
	cooling	2928	2928	680	680	1886	259
	Total discomfort hours	**2928**	**2928**	**680**	**680**	**1886**	**259**
First floor	overheating	447	432	410	247	35	325
	cooling	748	697	407	435	469	386
	Total discomfort hours	**1195**	**1129**	**817**	**682**	**504**	**711**
CONSUMPTIONS kWh/m²anno		**134**	**102**	**91**	**83**	**61**	**44**

Table 3. Discomfort hours. The total amount of hours in the hot season is equal to 2928 (1 June - 30 September)

4.1 Scheme B: Visible interventions

The first solution analyzed is the external insulation of the vertical envelope. We neglected the insulation of the internal side of the wall because several authors demonstrated that it determines the problem of surface and interstitial condensation (Stazi et al., 2009).

The study of discomfort hours (see table 3) on the ground floor during the summer, shows that the external insulation of the wall does not bring the temperature in the comfort range: all the hours in the hot season (equal to 2928) are hours of discomfort. This problem is well illustrated by the graph of operative temperatures (Fig. 8). This is due to the fact that there is great heat loss through the ground floor slabs which determines a very cold indoor environment. Therefore the adoption of one type of wall rather than another is of little importance.

Although slightly different, the figure show how, the retrofitting solution analyzed (scheme B) worsen the initial "as-built" condition (scheme A). In fact, insulation of the walls, by impeding the heat transfer from outdoors, further lowers the indoor temperatures and is therefore a disadvantage in terms of comfort.

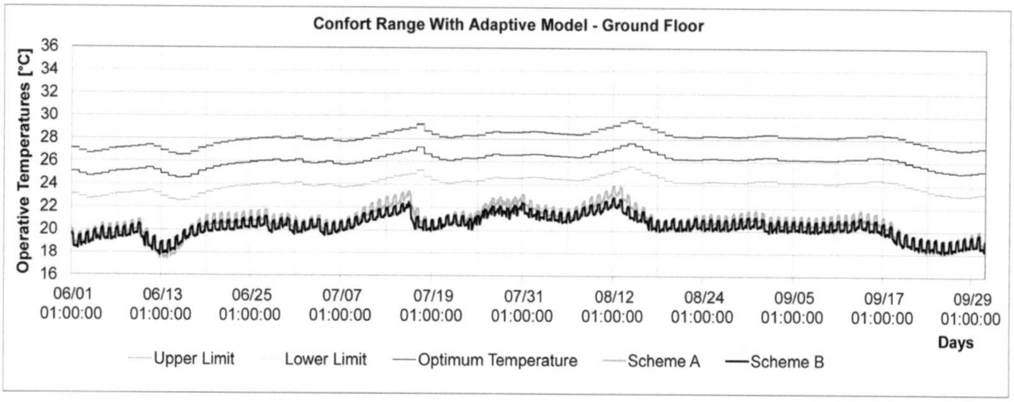

Fig. 8. Operative temperatures on the ground floor for scheme A ("as built" situation) and scheme B (external insulation of the vertical envelope)

A similar comparison was made on the first floor: the impact of external insulation of the walls was assessed, with no modifications being made to the other parts of the building.

In this case there is no problem of heat dispersion through the floor slabs and, together with the night time temperatures which are too low there is also the opposite problem of overheating during the daytime.

The graph of the operative temperatures (fig. 9) and the hours of discomfort (table 3) show that the insulated wall behaves better than the as built wall, with smaller fluctuations and temperatures which are closer to the comfort range. There is a reduction of both hours of nighttime discomfort as a result of cooling (from 447 to 432) and hours of daytime discomfort due to overheating (from 748 to 697).

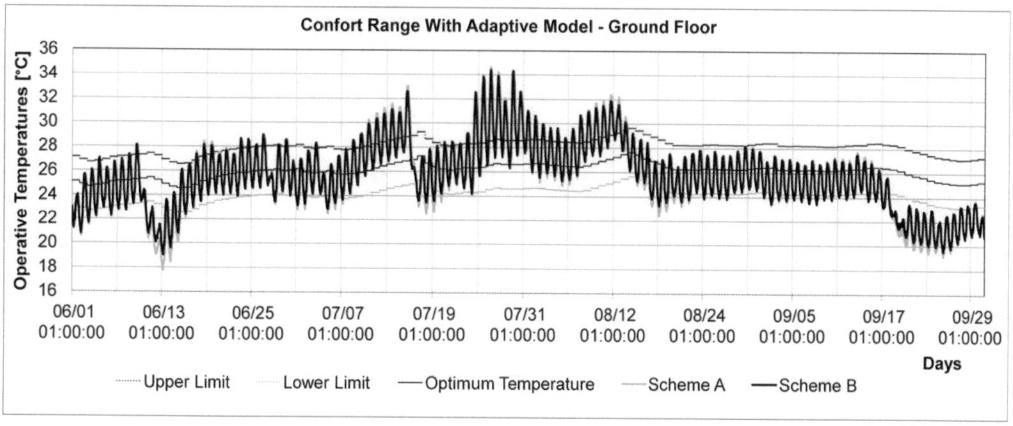

Fig. 9. Operative temperatures on the first floor for scheme A ("as built" situation) and scheme B (external insulation of the vertical envelope)

4.2 Scheme C: Not visible interventions

Interventions that do not affect the exterior image of the building regard the insulation of the ground floor slabs and intermediate floors (in relation to the fact that each floor has different apartment owners), insulation of the roof, introduction of double glazing.

From the figure showing the operative temperatures on the ground floor (Fig.10) it can be seen that the insulation of the ground floor slab and the introduction of double glazing determine a great reduction in the discomfort caused by the very low temperatures since this solution brings the temperature back within the comfort range (impact of 4 - 5°C on operative temperatures). This is clearly visible even from discomfort hours (see table 3) with a 77% reduction in dissatisfaction levels, from an initial value of 2928 to a value of 680 (the percentage is calculated on the total discomfort hours in the as built scheme=2928).

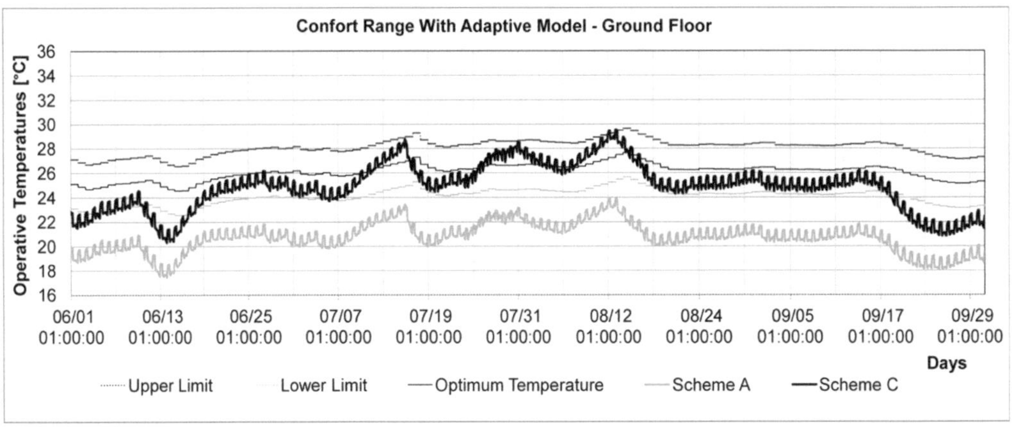

Fig. 10. Operative temperatures on the ground floor for scheme A ("as built" situation) and scheme C (insulation of the floors, insulation of the roof and introduction of double glazing)

The analysis of the operative temperatures (fig. 11) and discomfort hours (see table 3) on the first floor, shows that the not visible interventions determine a reduction of dissatisfaction

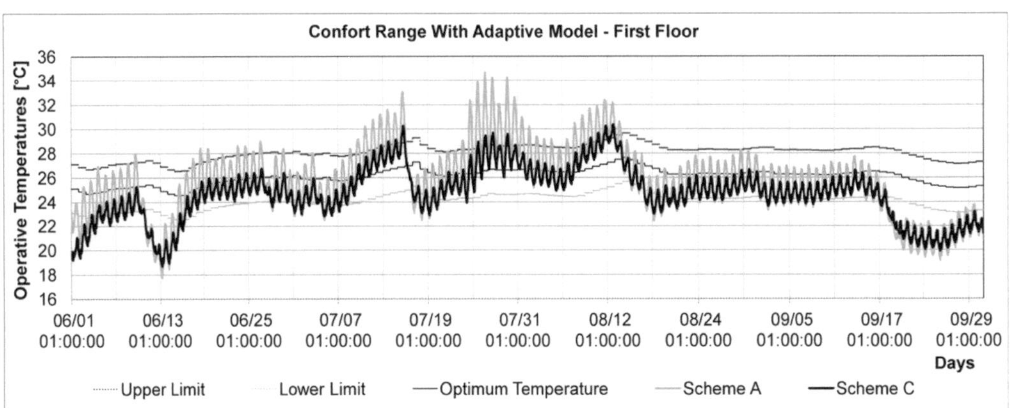

Fig. 11. Operative temperatures on the first floor for scheme A ("as built" situation) and scheme C (insulation of the floors, insulation of the roof and introduction of double glazing)

levels both for daytime overheating (discomfort hours from an initial value of 447 to a value of 410) and for nighttime cooling (discomfort hours from an initial value of 748 to a value of 407). So on the first floor there is a 30% decrease in the total hours of discomfort, from 1195 to 817 (the percentage is calculated on the total discomfort hours in the as built scheme=1195).

4.3 Scheme D and E: Combination of different interventions and comparison with a new lightweight building

Visible and not visible interventions were combined, considering an "optimized" building (scheme D= scheme B + scheme C). Said optimized scheme was compared with Scheme C (only interventions that are not visible) and with scheme E (building of new lightweight construction with low thermal transmittance).

The analysis of the operative temperatures (fig.12 and table 3) on the ground floor shows that the total discomfort hours in scheme D due to the cold environment increase substantially in relation to scheme C (from 680 to 1886) since the internal gains are very low and there is a great reduction of the thermal flows entering form the external envelope.

In the scheme E the hours of discomfort are low (259) if compared with other schemes. This scheme is slightly better because the most important problem on the ground floor is the discomfort for cooling and a wall with a low transmittance has to be preferred.

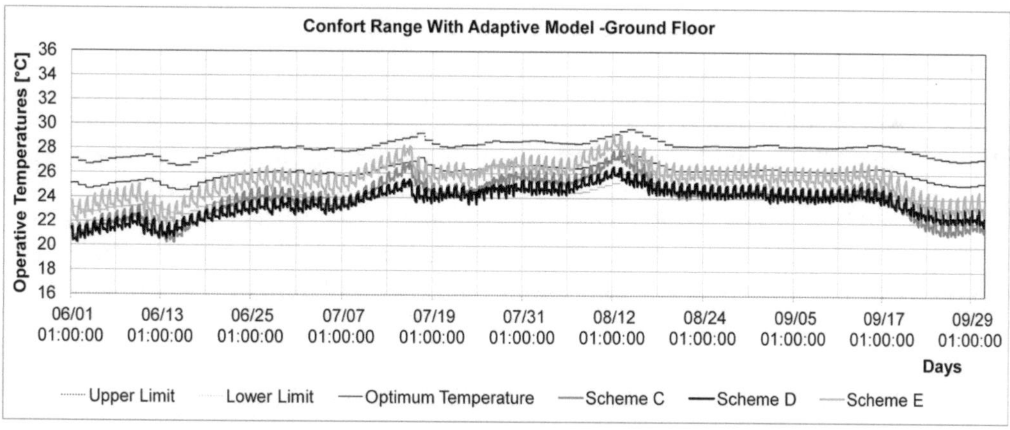

Fig. 12. Operative temperatures on the ground floor for scheme C (insulation of the floors, insulation of the roof and introduction of double glazing), scheme D (a combination of scheme B and C) and scheme E (new lightweight construction)

On the first floor together with the night time temperatures which are too low, there is also the opposite problem of overheating during the daytime.

In the scheme D the discomfort hours (fig. 13 and table3) for heating substantially decrease from 410 to 35 because the external insulation greatly reduces the thermal flows entering from the outside and the internal mass absorbs the internal loads. The total discomfort hours are 504, lower than in the scheme C.

The lightweight building (scheme E) presents some problem of daytime overheating. The discomfort hours of that scheme (with respect to the scheme D) increase up to a value of 325. This solution is slightly better than the scheme C.

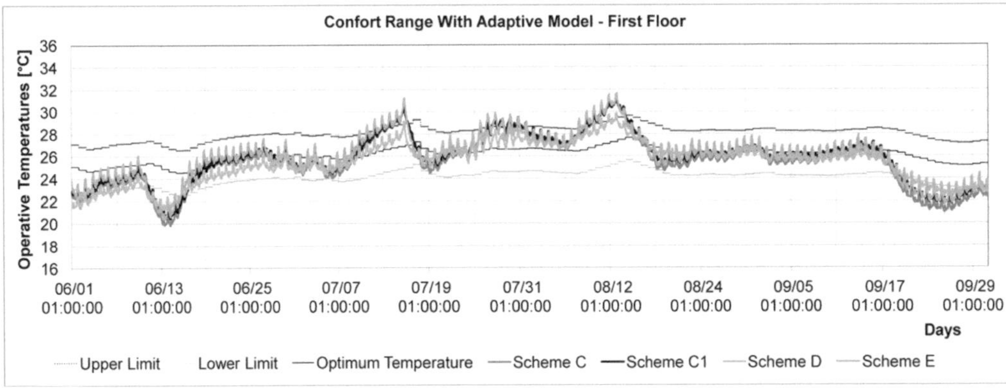

Fig. 13. Operative temperatures on the first floor for scheme C (insulation of the floors, insulation of the roof and introduction of double glazing), scheme D (a combination of scheme B and C) and scheme E (new lightweight construction)

4.4 The optimal configuration from the comfort point of view

The study shows that from the point of view of the summer comfort the most significant interventions are the thermal insulation of the floors and roof and the replacement of single panes with double glazing without intervening on the vertical walls (scheme C). As the matter of fact there is a great reduction of discomfort levels either with the scheme C (intervention on the not visible elements) or with the scheme D. In the first case the result show: in the ground floor a 77% reduction of discomfort hours; in the first floor a 30% reduction. In the scheme D the discomfort hours slightly decrease: 35% in the ground floor and 57% in the first floor. The scheme C present a total amount of discomfort hours of 1497 (680+817) while the scheme D present a total of 2393 (1886+507). Then it would be preferable to adopt the scheme C because it guarantees summer comfort and it does not alter the historic vertical walls.

The last phase of the study was to assess the impact of introducing in the scheme that proved to be the best (scheme C) an insulated ceiling (scheme C1). For the climatic zone of interest, the Italian standard imposes: a limit of 0.30 W/m^{2}K for the thermal transmittance of the horizontal attic floor (that delimits the heated area); a limit of 0.80 W/m^{2}K for the thermal transmittance of the superior pitched roof slab (separating an unheated room and the external environment).

The simulation (see table 3 and Fig.13) showed how the introduction of an insulated attic slab greatly reduces the hours of discomfort due to overheating (from 410 of scheme C to 247 of scheme C1).

This configuration resulted to be the optimal situation with respect to all the schemes studied.

4.5 Consumptions

We carried out a study on winter consumption to ensure that the identified optimal configuration does not lead to high winter consumptions (see table 3).

It resulted that the interventions on the elements that are not visible (scheme C) determine a reduction on winter consumption from an initial value of 134 to a value of 91 kWh/m^{2}year. The introduction of an insulated attic floor (scheme C1) reduces the winter consumption to 83 kWh/m^{2}year.

With the insulation of the external walls (scheme D) said value is further reduced to 61 kWh/m^{2}year.

Even if the exclusion of the intervention on the external walls is not congruent with the Italian regulations on energy saving (that for the external walls set a limit of thermal transmittance), a consumption value of 83 kWh/m^{2}year can be considered acceptable since it is slightly higher than the value obtained with the insulation of all structural components of the traditional building (61 kWh/m^{2}year).

A lightweight building realized according to standards for energy saving reduces the winter consumption to 44 kWh/m^{2}year.

5. Conclusion

An experimental and analytical study was carried out on a traditional farmhouse in central Italy. The study allowed the thermal behaviour of the building to be analysed according to different usage and weather conditions. It was also possible to identify the problems of summertime discomfort which, on the ground floor, are due to the considerable loss of heat through the floor slabs and, on the first floor, to the overheating of the rooms. In this regard retrofitting was proposed according to the new legislation on energy saving, and a comparison was made between various types of retrofit interventions carried out on the external walls in terms of both summer comfort and winter consumptions.

In particular (in consideration of the historical value of the stone walls) were distinguished the interventions on visible constructive elements (realization of an external insulation coating on the vertical walls) from interventions on the not visible elements (that do not alter the architectural image of the building: insulation of roof, insulation of intermediate and ground floors, introduction of double glazing).

Since discomfort due to cooling is of great importance, the insulation of the ground floor has a greater impact than the other interventions. The combination of said intervention with the insulation of the roof and the introduction of double glazing leads to a 77% decrease in discomfort hours. The insulation of the vertical walls with an external coating is too invasive for the image of the building and not very significant in terms of summer comfort and energy savings.

The intervention on the not visible elements has to be preferred since it determines adequate summer comfort and acceptable winter consumptions without altering the historic vertical walls.

The study also reports a comparison in terms of comfort and energy consumptions with a new building with low thermal capacity and low thermal transmittance.

The study could be useful to adjust the regulations on energy saving for traditional existing buildings of historic value in temperate climates.

6. References

C. Di Perna, F. Stazi, A. Ursini Casalena & M. D'Orazio, (2011) Influence of the internal inertia of the building envelope on summertime comfort in buildings with high internal heat loads. *Energy and Buildings* 43 (2011) 200-206

Dili A.S., Naseer M.A. & Zacharia Varghese T. (2010). Passive control methods of Kerala traditional architecture for a comfortable indoor environment: A comparative investigation during winter and summer. *Building and Environment* 45 (2010) 1134-1143

DS/EN 15251:2007, Indoor environmental input parameters for design and assessment of energy performance of buildings addressing indoor air quality, thermal environment , lighting and acoustics.

F. Stazi, C. Di Perna & P. Munafò, (2009) Durability of 20-year-old external insulation and assessment of various types of retrofitting to meet new energy regulations. *Energy and Buildings* 41 (2009) 721–731

Givoni, B. (1998). *Climate consideration in Building and Urban Design*, John Wiley and Sons, ISBN 0-471-29177-3, USA.

Grosso M. (1998). Design Guidelines and technical Solutions for Natural Ventilation, In: *Natural Ventilation in Buildings - a Design Handbook*, Allard F.; James & James (GBR), 1998, 195-254, ISBN: 1873936729

N. Cardinale, G. Rospi & A. Stazi (2010) Energy and microclimatic performance of restored hypogenous buildings in south Italy: The "Sassi" district of Matera. *Building and Environment* (2010) 94-106

R.J.Fuller, A. Zahnd & S. Thakuri, Improving comfort levels in a traditional high altitude Nepali house. *Building and Environment* 44 (2009) 479-489

Rijal H.B., Yoshida H. (2002). Comparison of summer and winter thermal environment in traditional vernacular houses in several areas of Nepal. *Advances in Building Technology* (2002)1359-1366

Ryozo O. (2002). Field study of sustainable indoor climate design of a Japanese traditional folk house in cold climate area. *Building and Environment* 37 (2002) 319-329

S. Martin, F. R. Mazarron & I. Canas, (2010) Study of thermal environment inside rural houses of Navapalos (Spain): The advantages of reuse buildings of high thermal inertia. *Construction and Building Materials* 24 (2010) 666-676

Wang F., Liu Y. (2002). Thermal environment of the courtyard style cave dwelling in winter. *Energy and Buildings* 34 (2002) 985-1001

Youngryel R., Seogcheol K. & Dowon L. (2009). The influence of wind flows on thermal comfort in the Daechung of a traditional Korean house. *Building and Environment* 44 (2009) 18-26

Z. Yilmaz, (2007) Evaluation of Energy efficient design strategies for different climatic zones: Comparison of thermal performance of buildings in temperate-humid and hot-dry climate, *Energy and Buildings* 39 (2007) 306-316

4

High Energy Performance with Transparent (Translucent) Envelopes

Luis Alonso, César Bedoya, Benito Lauret and Fernando Alonso
E.T.S.A.M. School of Architecture and School of Computing (UPM)
Spain

1. Introduction

Energy efficiency is coming to the forefront in the architecture, as, apart from the significance of a reduced environmental impact and increased comfort for users, the current energy crisis and economic recession has bumped up the importance of the financial cost of energy.

Since the Kyoto Protocol was signed in 1997, governments all over the world have been trying to reduce part of the CO2 emissions by tackling building "energy inefficiency". In Europe today, the tertiary and housing sectors account for 40.7% of the energy demand, and from 52 to 57% of this energy is spent on interior heating (Willems & Schild, 2008). The new world energy regulations, set out at the European level by the Commission of the European Communities in the First Assessment of National Energy Efficiency Action Plans as required by Directive 2006/32/EC on Energy End-Use Efficiency and Energy Services, (Commission of the European Communities, 2008) indirectly promote an increase in the thickness of outer walls, which, for centuries, have been the only way of properly insulating a building.

The use of vacuum insulation panel (VIP) systems in building aims to minimize the thickness of the building's outer skin while optimizing energy performance. The three types of vacuum chamber insulation systems (VIS) most commonly used in the construction industry today -metallized polymer multilayer film (MLF) or aluminium laminated film, double glazing and stainless steel sheet or plate (Willems & Schild, 2006) -, have weaknesses, such as the fragility of the outside protective skin, condensation inside the chamber, thermal bridges at the panel joints, and high cost, all of which have a bearing on on-site construction (Baetens et al., 2010).

Apart from overcoming these weaknesses and being a transparent system, the new F^2TE^3 (free-form, transparent, energy efficient envelope) system that we propose has two added values. The first is the possibility of generating a structural skin or self-supporting façade. The second is the possibility of designing free-form architectural skins. These are research lines that the Pritzker Architecture Prize winners Zaha Hadid, Frank Gehry, Rem Koolhaas, Herzog & de Meuron, among many other renowned architects, are now exploring and implementing.

To determine the feasibility of the new envelope system that we propose, we compiled, studied and ran laboratory tests on the materials and information provided by commercial brands. We compared this information to other independent research and scientific trials on VIPs, such as Annex39 (Simmler et al., 2005), and on improved core materials, such as hybrid aerogels and organically modified silica aerogels (Martín et al., 2008), conducted by independent laboratories like Zae Bayern in Germany (Heinemann et al., 2009), the Lawrence Berkley Laboratory at the University of California (Rubin and Lampert, 1982) or the Technical University of Denmark (Jensen et al., 2005).

After studying the results, we discovered valuable innovative ideas that we exploited to design the new high energy efficient envelope that should outperform the elements now on the market.

The remainder of the chapter is structured as follows. In Section 2 (Research) we explain the rationale of the epistemological study of the system, and how the experimental study combining computer simulations and empirical trials was run. In Section 3 (Experimental study) we describe the design of the proposed envelope system (F^2TE^3), explaining the solutions adopted in this new system and the improvements on other existing systems. Section 4 (Free-Form, High Energy Performance, Transparent Envelope System (F^2TE^3) Design) analyses how the F^2TE^3 system overcomes the weaknesses detected in existing VIP systems. Finally, Section 5 (Conclusion) discusses final conclusions.

2. Research

Today's architectural vanguard demands a building system such as is proposed in this research: a lightweight, variable geometry, seamless high energy performance system that also permits the passage of natural light and backlighting.

No system combing all these features exists as yet, and similar systems are not absolutely free form and translucent, are not seamless and/or have a very limited thermal response.

From the viewpoint of energy performance, of the three types of translucent insulations that there are on the market (plastic fibers, gas and aerogel), we found that the insulation that best meets the needs of the new system that we propose is aerogel (Ismail, 2008; Wong et al., 2007).

Aerogel and nanogel (granular silica gel) have four advantages for use as thermal insulation in translucent panels:

a. Transparency: aerogel is comparable to glass in terms of transparency (Moner-Girona et al., 2002). Monolithic aerogel light transparency can be as high as 87.6%, even greater than what some gas-insulated glasses can offer. Granular nanogel offers a greater translucency than any other traditional insulation material (including plastic fiber blankets).
b. Insulation: on top of transparency, it is an excellent insulator. According to published data (Baetensa et al., 2010), the thermal performance of a 70 mm nanogel-filled VIP is better than a 270 mm-thick hollow wall. The insulation values of a 15 mm nanogel-filled polycarbonate sheet are higher than any double glazing of similar thickness.
c. Lightness: aerogel can be three times as heavy as air (Rubin and Lampert, 1982), which means that the panels that employ this material for insulation are very lightweight even though the density of the aerogel that we use as insulation is 50-150kg/m^3.

d. Versatility: monolithic aerogel can be shaped as required. Being a nano-sized filling, granular nanogel can be used to fill any chamber.

Let us note that although there are many prototypes and patents (Gibson, 2009; Bartley-Cho, 2006) for the transparent and translucent high energy efficient aerogel-implemented façade panels under study, they are extremely difficult to analyze because there is not a lot of information available and it is not easy to get physical samples of these panels. For this reason, this research has focused on commercial products that are on the market. Most of these products use granular silica aerogel (nanogel).

In the following, we analyze these translucent and transparent panels, setting out their strengths and weaknesses and our findings as a result of this study.

2.1 Translucent systems

In this type of systems we have analyzed systems composed of granular silica gel-filled polycarbonate, reinforced polyester and double glazed vacuum insulated panels.

Nanogel-filled cellular polycarbonate panels are the most widespread system on the market. They have the following strengths and weaknesses:

Strengths: Thanks to its low density 1.2g/m^3, this is a very lightweight material. It has a high light transmission index ±90% (almost the transparency of methacrylate). It is a low-cost material for immediate use. And, at the competitiveness level, it is the least expensive envelope assembly.

Weaknesses: Durability is low. Most commercial brands guarantee their polycarbonate panels for only 10 years (as of when they start to deteriorate), whereas nanogel has a very high durability. These panels are very lightweight but very fragile to impact. Even though nanogel is an excellent acoustical insulator, the slimness of these panels means that they have acoustic shortcomings.

No more than two types of reinforced polyester panels are commercialized despite the potential of this material. They have the following strengths and weaknesses:

Strengths: Good mechanical properties: glass fiber reinforced polyester resin core composites offer excellent flexibility, compressibility and impact resistance. Good malleability: they could be shaped according to design needs but no existing system offers this option. Durability is good, as there are methods to lengthen the material's service life considerably (twice that of polycarbonate), like gelcoat coatings or protective solutions with an outer layer composed of a flexible "glass blanket".

Weaknesses: There is no self-supporting (structural) panel that is standardized and commercialized worldwide. Existing systems have design faults, as they include internal aluminum carpentry or substructures, whereas there is, thanks to the characteristics of reinforced polyester, potential for manufacturing a self-supporting panel (as in the case of single-hull pleasure boats). It is also questionable ecologically, as the polyester is reinforced with glass fiber, which has detracted from its use in building. However, this could change with the advent of new plastic and organic fibers and resins. Economically speaking, reinforced polyester manufacturing systems are very expensive, because either processes are not industrialized or, on the other hand, they rather technology intensive like, for example, pultrusion.

Double glazed vacuum insulated panels (VIP) are still at the prototype stage. Although research and prototypes abound (HILIT+ y ZAE BAYERN, for example) (Fricke, 2005), there are only a couple of commercial brands:

Strengths: Thanks to the combination of vacuum and aerogel insulation (both monolithic and granular), they provide the slimmest and best insulation system in the building world (0.5W/m^{2}K). Transparency levels for some prototypes using monolithic aerogel are as high as 85% for thicknesses of 15 to 20 mm. Additionally, the service life of the glazing and the gel is very similar.

Weaknesses: Product of a combination of vacuum core and double glazing, this component is fragile, especially prone to impact-induced breakages. The high cost of molding glass into complex geometries rules out its use as a free form system. It is a system that depends on substructures and other components for use (Figure 1).

Fig. 1. Brittle fracture of panels tested in the laboratory of the Department of Construction and Technology in Architecture, School of Architecture. (UPM.) E.T.S.A.M.

Findings: After a comparative analysis of over one hundred and forty seven (147) commercial products, and the detailed evaluation of the best eight (8) (Figure 2), we can confirm that fiber reinforced polyester resin panels perform better than the best polycarbonate panels. But these improvements are unable to offset their high production and environmental costs, generating commercially uncompetitive products.

These panels have two unexploited design lines, such as adaptation to new less harmful natural cellulose resins and fibers, the design of insulation for variable geometry translucent skins, or structural improvement for use as a self-supporting component.

As regards energy efficiency, these products offer more energy-saving performance than a 27cm thick traditional wall. With only 7 cm, they have an U value of 0.28W/m^{2}K, which amounts to 7% better energy performance compared to a traditional wall and with 4% less thickness.

Looking at double glazed VIP panels; the data indicate that, although still at the prototype stage, panels like these are the best commercial solution, as they offer the best thermal and acoustical insulation performance and optimal light transmission.

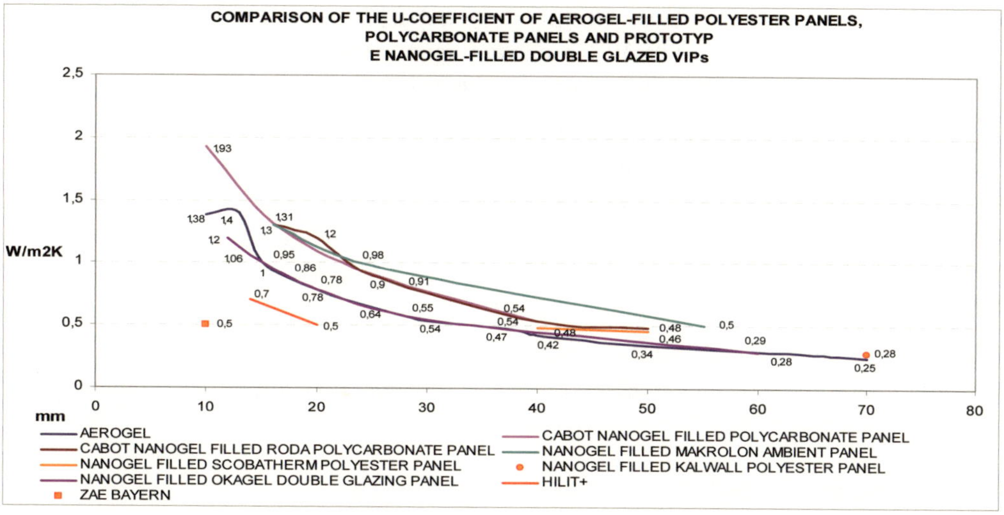

Fig. 2. Comparison of the U-value of the most significant commercial systems and prototypes implemented with an aerogel core existing on the market.

At the acoustical and thermal level, the VIP panel is the best of the envelopes examined, as a 15 mm panel insulates equally as well as a traditional mass wall in terms of energy expenditure (saving). We find that, with a thickness of just 60 mm, this product improves the energy efficiency performance of a 271.5 mm cavity wall.

2.2 Transparent systems

All panels implemented with aerogel instead of nanogel are transparent. They have a high solar transmittance and low U value. At present all these systems are non-commercial prototypes, about which little is known. Noteworthy are two aerogel-insulated double-glazed vacuum insulation panels (VIP):

a. 4-13.5-4 / 21.5 mm double-glazed vacuum panels filled with monolithic aerogel with a pressure of 100hPa1 in the aerogel chamber. The heat transfer coefficient Ug has a U value of 0.7 W/m²K for 14 mm and 0.5 W/m²K for 20 mm compared to the 1.2W/m²K offered by 24 mm commercial nanogel-filled double glazing VIP. This almost doubles the insulation performance of the best commercial translucent panel. Light transmission depends on the angle of incidence, but varies from 64.7 to 87.5%. The sound attenuation index is 33dB for a panel thickness of 23 mm and noise reduction is expected to be improved to 37 dB. The energy saving compared with a dwelling that is glazed with gas-insulated triple glazing (argon and krypton) is from 10% to 20% greater.
b. 10 mm double-glazed vacuum insulation panels with aerogel spacers inside the core (unlikely to be commercialized for another two or three years). The heat transfer coefficient Ug for 10 mm panels has a U value of 0.5 W/m²K. This is the best of all the panels studied so far, where light transmission is equal to glass.

Findings: From the analysis of the transparent panels, VIPs unquestionably come the closest to what we are looking for in this research. The only arguments against VIPs are that they are at the prototype stage. This means that they are not on the market, nor have they been tested, approved or industrialized. All this has an impact on cost. Also being the product of evacuating double glazing panels, VIPs are very fragile. Versatility is limited because, owing to the panel generation process, the maximum dimensions to date are 55 x 55 cm.

From the materials technology analysis, we find that transparent monolithic aerogel-insulated VIPs are the material that best conforms to the goals of transparency, insulation and lightness, provided that we accept that the panels are fragile, non-commercialized prototypes and are not very versatile in terms of size.

3. Experimental study

Following up the results of the theoretical study outlined in Section 2, we now compare these findings with the results of an empirical experiment and computer-simulations of the real commercial panels to which we had access.

3.1 Computer simulation

Existing programs in the market for calculating the energy efficiency are very focused in the pursuit of thermal comfort inside the building. Therefore, we have had to adapt this program's calculating tools to finish getting results as realistic as possible.

On one side we have had modeled the environment and external environmental conditions of the test site: For this we have used the additional tools of Autodesk calculation programs, such as the Solar Tool (Figure 3) that parameterized the values of sun exposure during the test period, depending on the latitude and longitude of the testing place, or the Weather Tool that transforms all the atmospheric data (wind, temperature, rainfall, percentage of clouds, etc.) in to numerical parameters for use them in the thermal computation.

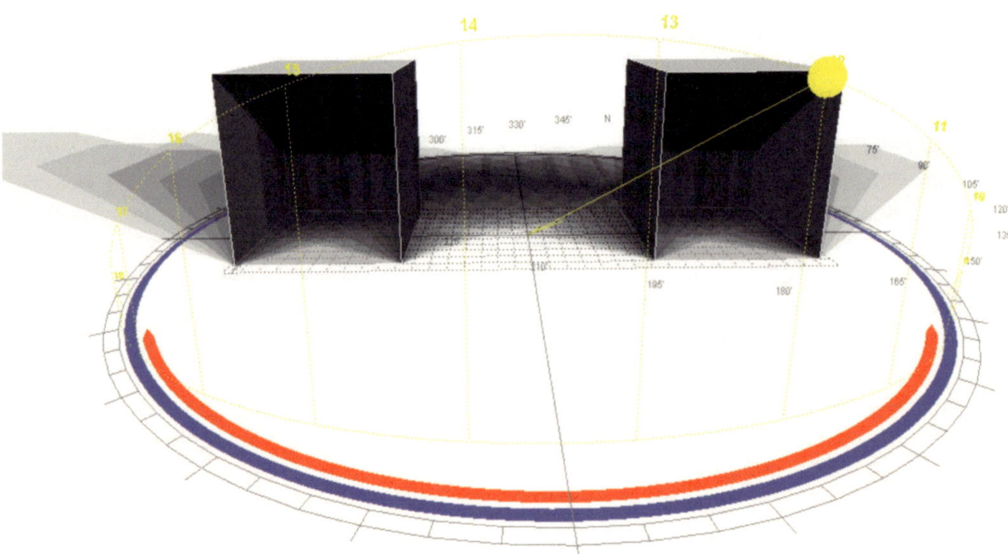

Fig. 3. Computer simulation: Solar data of March 14, 2010 at 12:00 in Madrid. (Latitude: 40 ° 23′ North, Longitude: 4 ° 01′ West)

On the other hand we has worked with a small-scale models, modelling the real test box, that is a cube of 60x60x60 cm, when such programs are more used to calculate large volumes built, with the adjustments that this has brought. We also have had to model all the panels systems that we tested, with the materials that are composed, in order to include them into our material library.

This whole process of modeling has been done trying to rely on independent test data. We do not want to use the data provided by trademarks, because after all, are the data that we want to verify objectively.

These programs are designed for regulatory compliance (UK and USA, among other). So the programs are oriented to guarantee thermal comfort, through adding contributions from outside air, and even turning on some air conditioning systems without the programmer's requests. Therefore, in order to get actual results, we had to overcome programming obstacles, disabling predetermined functions in some of the programs.

We used two powerful programs for the computer calculation: DesignBuilder and Ecotect. Although we have worked with both programs at the calculation, at the end we decide using DesignBuilder at the final testing, mainly due to the support the Ministry of Energy of the United States offered to the program and because the data that offered by this program correspond more to those data obtained with trials.

Because of the shortage of information about aerogel and the impossibility of acquiring a sample, we decided to use the Design Builder program to conduct a trial by computer simulation under the same environmental conditions as the empirical trials run on the other panels. Figure 2 describes the behavior of a 25 mm aerogel sheet. We find that the test space has a uniform inside temperature of between 18 and 37 °C.

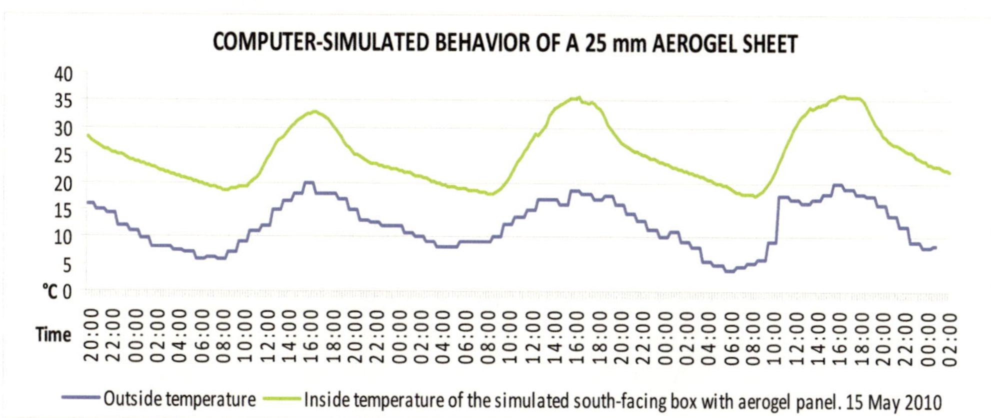

Fig. 4. Computer simulation for a 25mm thick sheet of silica aerogel with density 50-150kg/m^3 over 78 hours.

Figure 4 describes the behavior of a 25 mm Aerogel panel. We checked that the temperature inside the test space is evenly maintained between 18 and 37 °C. The curves at the loss periods are stretched, which means that this is a good thermal insulation, as it loses energy at a very long term and gradually.

3.2 Empirical trials

The trials are designed especially to examine the energy performance of the material. These trials were run at the Department of Building and Architectural Technology of the UPM's School of Architecture on boxes with an inner volume of 60 x 60 x 60 cm, insulated with 20 cm of expanded polyurethane. One of the box faces is left open by way of a window. The study elements are placed in this opening using a specially insulated frame. The trial involves exposing two such boxes to a real outside environment to study their behavior. The two boxes have two different windows: one is fitted with 6+8+6 double glazing with known properties as a contrast element and the other is fitted with the panel that we want to study. Data loggers are placed inside each box for monitoring purposes to measure and compare their inside temperature. The boxes are also fitted with a thermal sensor on the outside to capture the temperature to which they are exposed. The boxes are set in a south-facing position as this is the sunniest exposure (Figure 5).

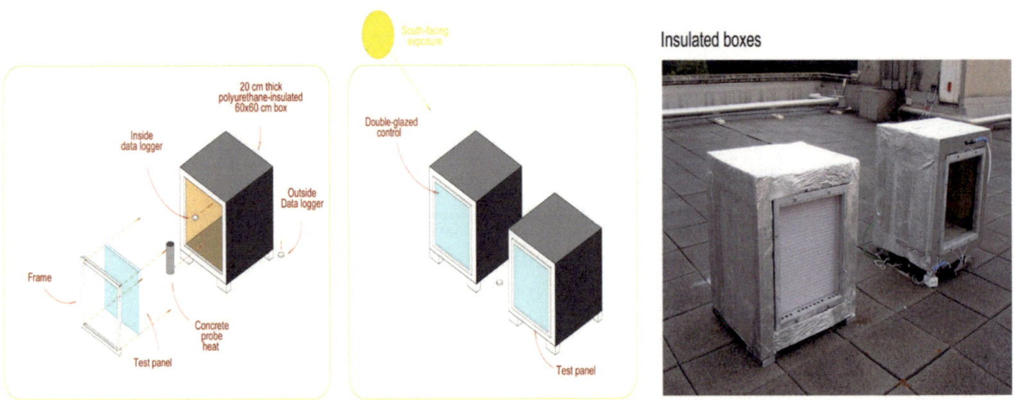

Fig. 5. Energy performance testing of the UPM system based on the hot box method

We ran twenty-eight (28) temperature-measuring trials using this system, and compared the performance of different thicknesses of commercial panels with 6+8+6 double glazing. Four (4) of these panels deserve a special mention.

3.2.1 Trial-1: 16mm nanogel-filled Cabot Lexan Thermoclear polycarbonate sheets (PC sheets)

Results: There is on average a 2.5 °C improvement in thermal properties over the commercial Climalit double glazing (which is 5 mm thicker than the panel) at night, and it insulates almost 6°C more than double glazing exposed direct solar radiation (Figure 6).

The fast enough rise of the curve at time of the thermal capture (day) is very similar in the two samples tested and compared (in the study panel and in the double glazing window sample probe), which means that their qualities are very similar solar collection, but there is a delay at the temperature increase with a difference of 24 minutes for the polycarbonate panel filled with nanogel, between one rise in the temperature and the other (between days). This confirms the market values in which, although showing a similar response of the double glazing, the polycarbonate panels filled with nanogel of similar thickness, usually have a small improvement if we compared them with a double glazing window.

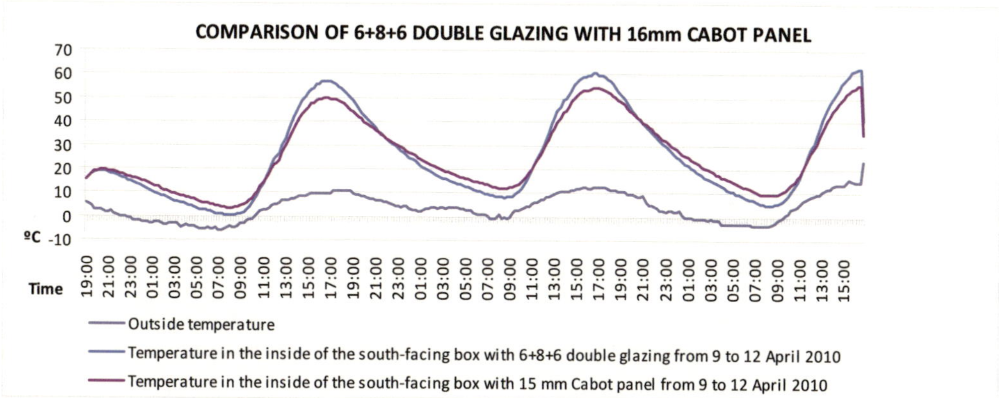

Fig. 6. Trial 1. Sample tested: (600x600mm of side) Policarbonate Panel of 16mm thick filled with nanogel. The test was made from the 9th (19:00 h) to 12 (15:00 h) in April 2010. At the E.T.S.A.M. School of Architecture & School of Computing (UPM) Spain

With the night loss, the difference between items is growing. The loss curve of the plate that is filled with nanogel is much more horizontal (which shows a U-value lower) than the loss curve of the sample window, leading to reduce the energy loss of the panel, and keeping the heat within the interior volume for more than 75 minuts and with a temperature that never is down the 10 °C although the outside temperatures was in the order of -4 °C.

These results support that a 16 mm polycarbonate panel filled with nanogel offers some benefits (although similar) than those offered by the double glazing of 20 mm (6+8+6).

3.2.2 Trial-2: 25mm nanogel-filled Cabot Lexan Thermoclear (triple-wall) polycarbonate sheets (PC sheets)

Trial-2: 25mm nanogel-filled Cabot Lexan Thermoclear (triple-wall) polycarbonate sheets (PC sheets).

Results: There is on average a 2 to 3 °C improvement in thermal properties over the commercial Climalit double glazing (which is 5 mm thinner than the panel) at night, and it insulates 15 to 20 °C more than double glazing exposed to direct solar radiation (Figure 7).

The solar collection curves are very different in this graph (figure 7). While in the double glazing, the rise of the curves is very fast and in 150 minutes (two hours) the curve has risen to nearly its maximum inside temperature, the panel filled with nanogel creates a flatter curve that marks a more homogeneous rise, and it takes 300 minutes (five hours) to reach its maximum inside value. Furthermore, there is a heating delay of 120 minutes for the box where we are testing the polycarbonate sheet filled with nanogel. All of this clearly shows the best performance of this panel versus the double glazing in daytime solar collection.

The loss curves at night make a similar trend, a difference of 360 minutes, six-hour delay between the thermal losses of 25 mm panel filled with nanogel and the double glazing window. A flat graph denotes a small U-value. Also noteworthy is that the graph of the polycarbonate panel filled with nanogel oscillates in a strip of between 25 °C and 40 °C as

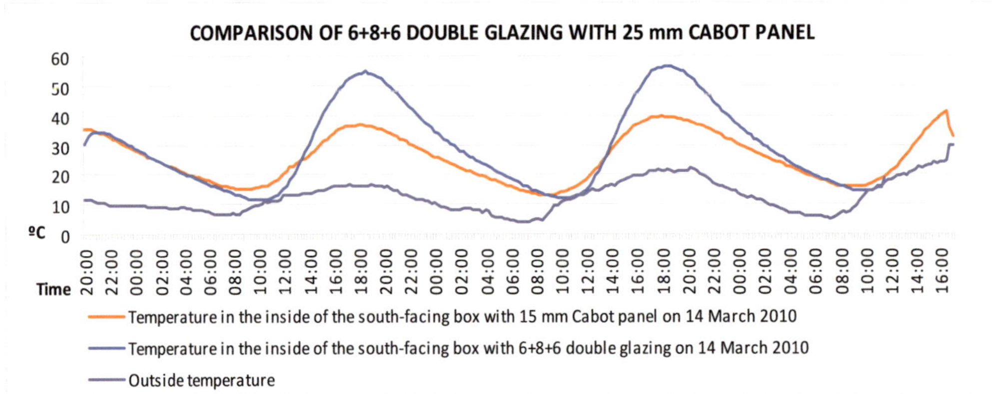

Fig. 7. Trial 2. Sample tested: (600x600mm of side) Policarbonate Panel of 25mm thick filled with nanogel. The test was made from the 14th (20:00 h) to 17 (16:00 h) in May 2010. At the E.T.S.A.M. School of Architecture & School of Computing (UPM) Spain

the maximum internal temperature inside the box, whereas the graph of the double glazing is between 55 ºC and 10 ºC.

These data show that increasing the insulation thickness leads a heat improvement of the system that is almost directly proportional to the thickness. With a increase of the 36% of the thickness, we obtained a thermal efficiency of 33% (20 °C improvement) in the solar collection and (5 ºC improvement) in night losses.

3.2.3 Trial-3: Bayer Makrolon Ambient S2S-25 sheet. 25 mm nanogel-filled twin-wall polycarbonate panel

Results: Behavior is very uniform. We get a 3 ºC improvement in thermal properties over the commercial Climalit double glazing, which is 5 mm thinner, at night, and it insulates 5 ºC more than double glazing exposed to direct solar radiation (Figure 8).

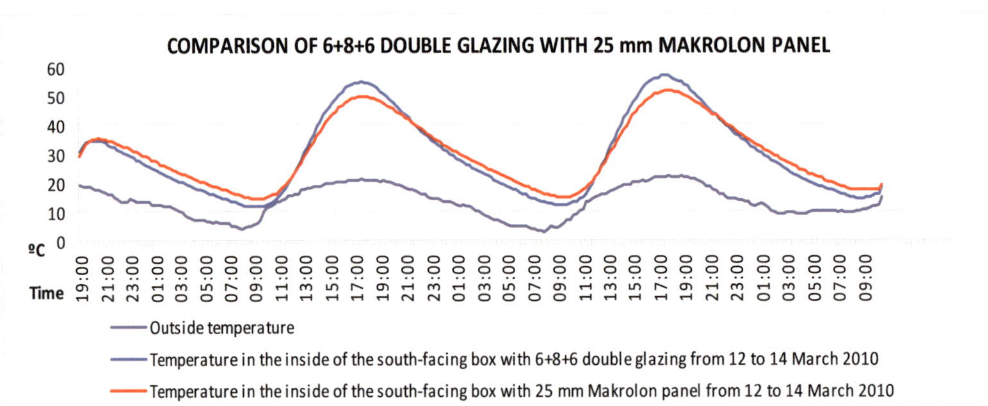

Fig. 8. Trial 3. Sample tested: (600x600mm of side) Policarbonate Panel of 25mm thick filled with nanogel. The test was made from the 12th (20:00 h) to 14 (16:00 h) in March 2010. At the E.T.S.A.M. School of Architecture & School of Computing (UPM) Spain

The solar collection curves (day) are very similar in both the study panel as in the double glazing sample and ascend quickly. Even though that thermal uptake quality are very similar in both elements, there is a delay difference of the temperature increase inside the box where the polycarbonate panel is tested of 45 minutes over the double glazing window.

At night loss the panel filled with nanogel is more flat from the start, marking quick differences with the double glazing. This trend lasts all night which denotes a small U-value and keeps the heat within the interior volume of the box over 80 minutes.

This confirms the market values, which showed that despite having the same thickness of panel (25 mm) and use the same Cabot nanogel, there are big difference between the Makrolon panel and the Lexan Thermoclear panel.

3.2.4 Trial-4: 70 mm Okagel Okalux VIP Panel: Nanogel-filled vacuum insulation panel

Results: Temperature is homogeneous ranging between 17 to 32°C, and there is an almost constant difference of from 3 to 10°C compared with double glazing (Figure 9).

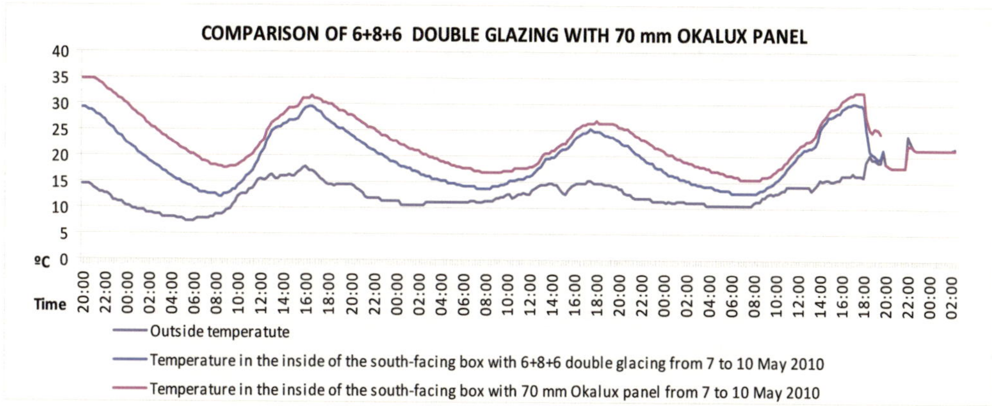

Fig. 9. Trial 4. Sample tested: (600x600mm of side) double glazing Panel/window of 70mm thick filled with nanogel. The test was made from the 7th (20:00 h) to 10 (16:00 h) in March 2010. At the E.T.S.A.M. School of Architecture & School of Computing (UPM) Spain

These four trials were evaluated and compared with the computer-simulated aerogel data (Figure 10) and data from the theoretical study. We found that, like the data output by the theoretical study, the real trials suggest that the materials behavior is suitable for designing the new envelope system. The very flat loss curves in the plot describe a very low U value. In terms of capture, there is a thermal difference of almost 30°C between the Okagel (VIP) panel and the worst of the tested panels. The difference between Okagel and the best-performing panel is almost 10 °C in terms of loss and capture. We have confirmed the experimental datum that likens the behavior of the Okagel panel to that of the computer-simulated aerogel.

From our computer-simulated experimental study, the data on organic aerogels supplied by the CSIC and the University of Barcelona, and the data from trials run at the University of Denmark on envelopes implemented with monolithic silica gels and the empirical trials conducted in this research, we arrive at the following conclusions:

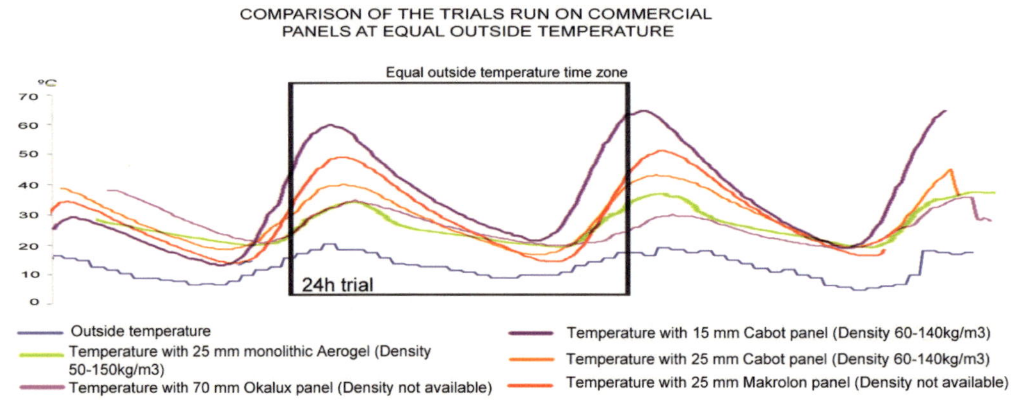

Fig. 10. Comparison of empirical data of commercial systems with computer simulation of 25 mm thick sheet of silica aerogel with density 50-150kg/m³ over 78 hours (temperatures inside the test boxes)

The best-performing panel from the energy saving viewpoint is 70 mm Okalux that has thermal differences with respect to the other panels ranging from 5 to 20°C. Also striking is the disparity in the results of the 25 mm and 15 mm Cabot panels with thermal differences of 15°C.

Although the specific temperatures and factors on each test day differed from one trial to another, the 70 mm nanogel-filled Okalux VIP panel performed similarly to the 25 mm aerogel sheet.

These are key data that are useful for designing a new lightweight, slim, high energy efficient, light-transmitting envelope system, providing for seamless, free-form designs for use in architectural projects.

4. Free-Form, High Energy Performance, Transparent Envelope System (F^2TE^3) design

The proposed component has better properties than double glazing VIP and reinforced polyester panels (Okalux and Kallwall) separately, as it combines the properties of the two to generate a new system and solve the problems specified earlier.

This section will present the "Proposal for a Free-Form, Transparent, Energy Efficient Envelope System (F^2TE^3)", and will explains the details of the new system, materials of construction, its size and overall value. Also shows the way to manufacture our system and the On-site assembly of the new system. We make a special reference of how the F^2TE^3 response to the weaknesses of VIP systems. Finally, we study the tests to determine the level of energy efficiency by calculation by computer.

4.1 Proposal for a Free-Form, Transparent, Energy Efficient Envelope System (F^2TE^3)

We propose a free-form design envelope system fabricated with cellulose fibers and polyester resin (or acrylic-based organic resin), and a vacuum core insulated with monolithic aerogel at a pressure of 1hPa. Being a self-supporting component, the system can perform structural functions, and seams between panels are concealed by an outer coating applied in situ (Figure 11).

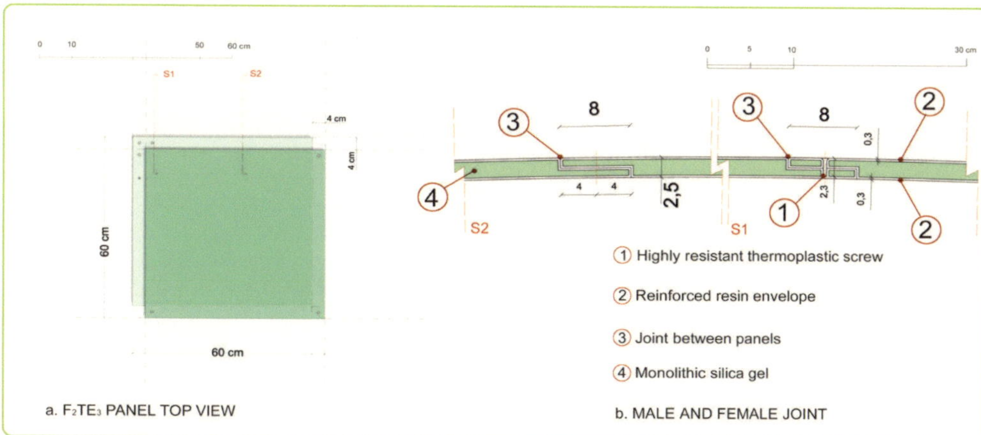

Fig. 11. F^2TE^3 Envelope Profile Design

Materials: The optimal materials are: (a) a natural cellulose fiber-reinforced epoxy resin matrix with similar performance to E-type glass fiber, with an outer gelcoat coating to protect it from external agents for the outer skin, on which the panel's resistance, protection and variable geometry depends, and (b) a thermal/acoustical insulation component composed of a monolithic silica gel-filled vacuum chamber.

Dimensions:

Based on tests run at the University of Denmark, we have to take into account the sol-gel process drying times required for the monolithic silica gel to generate a crystalline structure, the percentage of breakages due to size and, above all, the fact existing autoclaves are able to generate monolithic gel pieces no larger than 55 x 55 cm. The panel sizes will be 60 x 60 cm (length/width), and panel thickness will depend on whether the structure is self-supporting or a simple envelope. For modeling purposes, however, we have studied a 25mm thick panel, composed of two sheets of 3 mm thick reinforced resin and a vacuum core filled with monolithic silica gel (Figure 12).

The weight per unit of surface area will be from 15 to 7 kg/m2. Although dimensions could vary, the sheet width will be no greater than 600 x 600 mm and the minimum admissible flexion radius will be approx. 4000 mm.

Specifications:

Light transmittance, τD65: from 59% to 85% approx.
UV absorption: approx. 20%
Total energy: approx. 61%
Horizontal and vertical U-value: 0.50 W/m2 K
Thermal conductivity coefficient: 0.01 W/m K, estimated
Possible heat-and humidity-induced dilation: 3 mm/m approx.
Operating temperature: -70 to +80 °C
Weighted sound reduction value: estimated at 26-45 dB

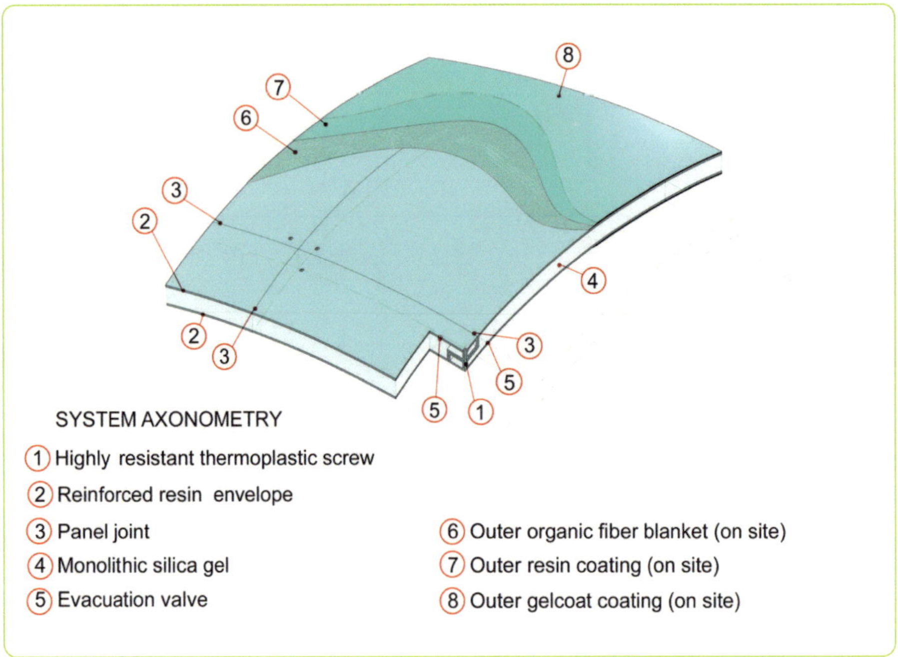

Fig. 12. F^2TE^3 system axonometry

4.2 System specifications

The F^2TE^3 is composed of the dry-seal connection of previously designed male and female edged panels (two female sides and two male sides on each panel that fit together seamlessly). Once the construction is in place, it is given an outer coating of fibers and resins and finally a gelcoat coating to protect the assembly from external agents.

Manufacturing and on-site assembly (Figure 13):

The off-site manufacturing process is composed of the following phases:

a. An easy-to-use, reusable and ecological molding process. As this does not have to be a structural mold, it can be made of compressed and sanded sheeting arranged according to the panel design.
b. The same mold is used to generate the two panel ends and its walls.
c. The walls are adhered to one of the ends using resin, and the assembly is filled with silica gel, where the panel itself acts as a mold to shape the monolithic silica gel inside the chamber.
d. Sol-gel technology is used to generate the aerogel inside the panel using an autoclave.
e. The top end is adhered to the panel using resins.
f. The panel is evacuated to a pressure of 1hPa inside the chamber.
g. The panels are transported separately to the site. Scaffolding and props are then used to assemble in situ the panels making up the façade. The panels are screwed together using highly resistant transparent thermoplastic screws.

The system should be manufactured under adequate health and safety conditions.

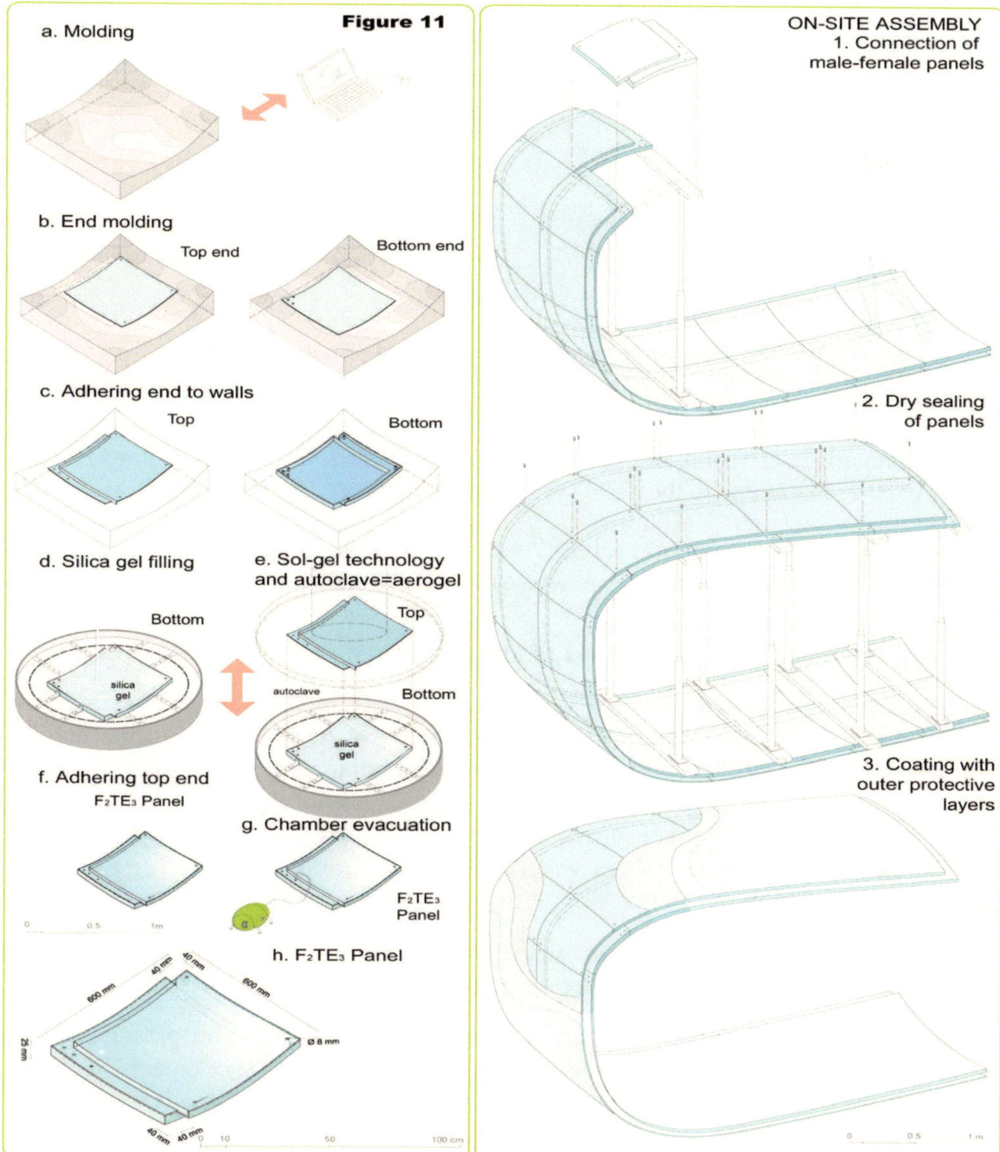

Fig. 13. Manufacturing and On-site assembly

4.3 Testing

A F^2TE^3 system with a thickness of 25 mm has been computer simulated to examine its energy-saving behavior compared with a computer-simulated aerogel envelope of the same thickness (Figure 14).

As shown in Figure 12, although the plots are displaced, the F^2TE^3 system returns a result very close to what would be achieved with monolithic aerogel without a barrier envelope (not feasible due to aerogel hydroscopy). Even with a barrier envelope, F^2TE^3 performance

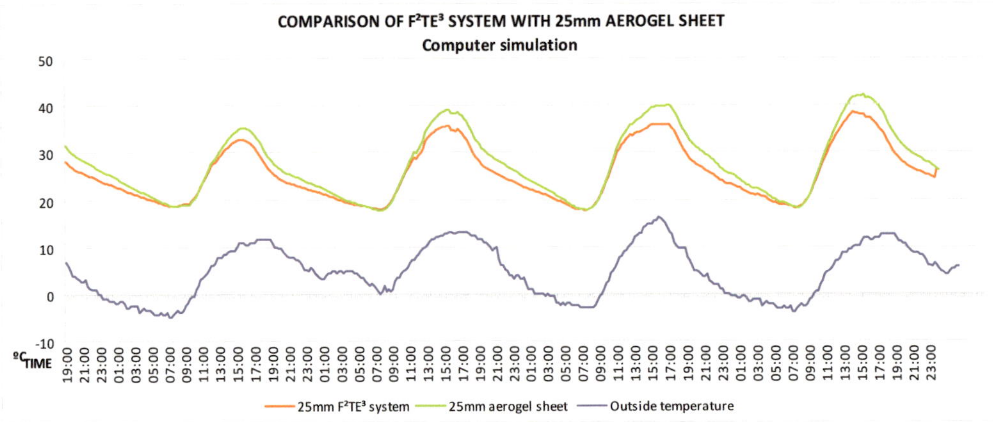

Fig. 14. Comparison of a computer simulation of a 25mm thick sheet of silica aerogel with a density of 50-150 kg/m³ with the F²TE³ system over 96 hours.

almost equals aerogel in terms of heat loss, with a very similar flat curve, where the U value is very small, but results for capture are worse at over 5ºC higher.

4.4 F²TE³ response to the weaknesses of VIP systems

Most studies conducted in the field of VIP elements (VIS) (Mukhopadhy et al., 2011) determine that there are four key obstacles to the use of these systems in the building industry: a) fragility of the outer skin, b) thermal bridging at panel edges and joints, c) vacuum chamber moisture permeation, and d) price. On top of these weaknesses, there is the added obstacle of a single element having to provide transparency and energy efficiency, plus a new demand from vanguard architecture for generating free forms by means of monocoque systems.

Theoretically, the new F²TE³ overcomes all the above objections raised against VIPs, offering the added values of transparency and free-form design:

a. Fragility of the outer skin: As the new system has to be a transparent and self-supporting structure, we have replaced the fragile outer barrier elements, such as a metallized polymer multilayer outer skin, double glazing and thin stainless steel sheets, with an element composed of highly resistant reinforced fibre resin.
b. Thermal bridge at panel edges and joints: Commercial panels are not designed to be assembled to form seamless elements. The new system that we propose is purposely designed to generate monocoque elements through a system of male and female panels. This system eliminates the thermal bridge between the panels (see Figure 15).
c. Moisture in the vacuum chamber: One of the key problems of VIP systems is condensation forming inside the chamber. Although some studies estimate that the hydroscopic silica smoke is capable of absorbing most of the minimal amount of condensation that forms without compromising its load-bearing or thermal capacities, the fact is that humidity inside the chamber ends up turning the monolithic aerogel opaque.
F²TE³ solves this problem by modifying the barrier thickness. The outside face is 3 mm thick compared with the inside face that should have thicknesses ranging from at least 6

to 8 mm to prevent condensation forming inside the chamber. We used the COAG's memorias2 program to plot the condensation graph of the new envelope, and the resulting sections are shown in Figure 16.
F^2TE^3 has a water vapour permeability value of 0,3 g/m²d (for 3mm thickness) less than some commercial VIP panels (0,5 g/m²d).

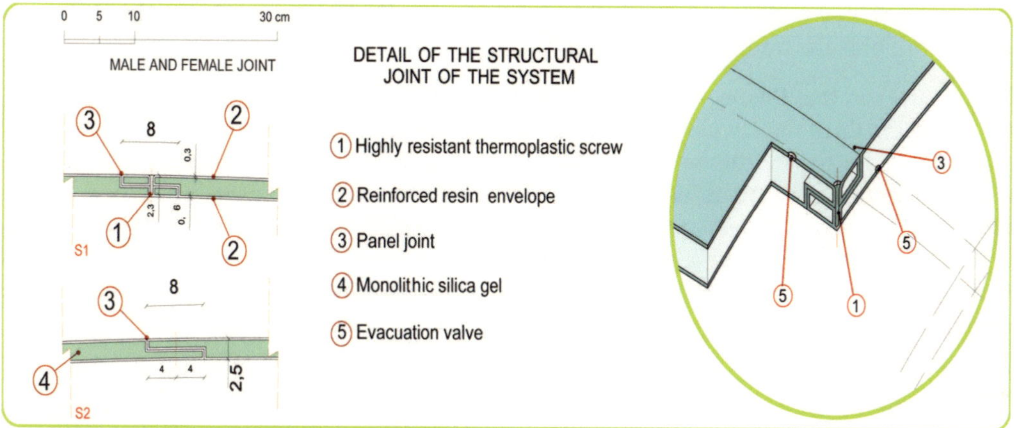

Fig. 15. Detail of the male and female joint of the F2TE3 system.

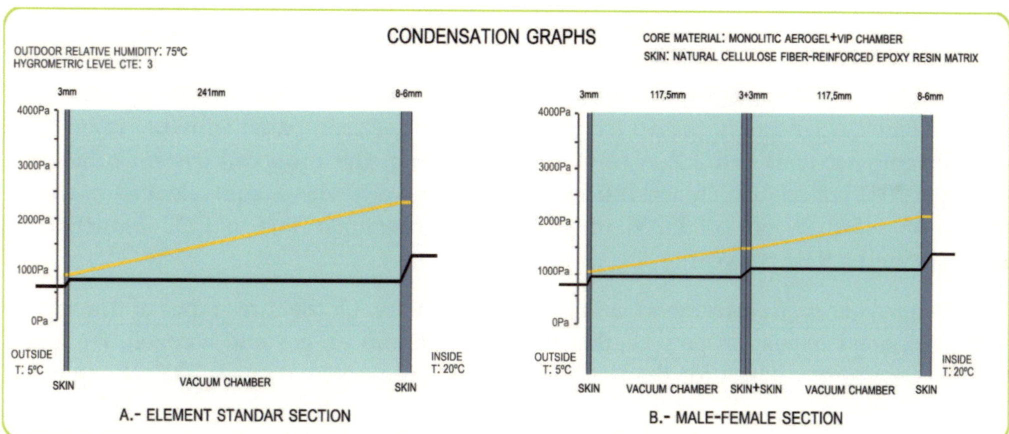

Fig. 16. Condensation graphs for a standard section and a male female section of the new F2TE3 system.

d. Price: VIP systems are now much higher cost than traditional building systems (approximately 200 €/m²). However, Technical University of Denmark (DTU) studies (Jensen et al., 2005) show that the mass produced product, using a fully automatic 10 m³ autoclave (as the sol-gel process is the most expensive part of the fabrication of this system), could be commercialized for less price.
For the purposes of our calculations, we will take a 150 m² detached dwelling and assume that the F^2TE^3 system costs 300 €/m² (If we use an conventional autoclave of 600 ml capacity (Parr Instrument Company, Moline, IL, USA) from the Air Glass Laboratory). On the other hand, the mean cost of a 6-12-6 double glazing aluminium

window with a thermal break is 360 €/m², whereas a plastered, double hollow brick lined, 4 cm thermally insulated, 6 inch rough brick conventional wall with single-layer finish costs on average 85 €/m². Assuming that the mean percentage of glazing is 45%, the current cost of a conventional envelope including the façade glazing is 162 €/m².

The surface area of the envelope for a 150 m² dwelling is 375 m². The cost of building a conventional wall will be 60,750 €, whereas F²TE³ costs 112,500 €, meaning that F2TE3 costs just over double the traditional system at this stage.

Note, however, that the net internal area of the dwelling built using the F²TE³ system is 148 m² (at 2.5 cm thick the envelope only takes up 0.833% of the gross external area), compared with 135 m² for the traditional system (at 30cm thick it takes up 9.7% of the gross external area). Therefore, developers using the new F²TE³ system can commercialize an additional net internal area of 13 m². Assuming that the final sale price per square metre of net internal area is 4,000 €, there will an additional gain of 53,560 €. This, deducted from 51,750 €, which is the result of subtracting the price of the conventional building skin from the cost of the proposed envelope, results in a net profit of 1,810 € for developers using F²TE³.

Clearly, this calculation does not take into account either the potential saving thanks to the use of the new system as a structural element (which could amount to a 20% saving over the total cost of executing the building works) or the potential time and labour saving using this new building system compared with the traditional envelope.

According to DIN standard 4701, "there would be an annual saving of from 0.9-1.3 l of oil or 1.0-1.5 m³ of gas per m² of naturally lit surface area, leading to U-value reductions of 0.1 W/m²K". The F²TE³ surface area would be the sum of the walled areas (168 m²) and glazed areas (206 m²). Accordingly, with a U-value of 0.35 W/m²K, compared with 0.44 W/m²K for the conventional wall and 1.2 W/m²K for the glazing, the proposed system achieves a saving of 2083 l of oil (0.7€/l) and 2403m³ of gas (0,55€/ m³). This is equivalent to an annual saving of 1458€ in oil or 1322€ in gas for the consumer and a CO_2 reduction of approximately 4.9 kg per year.

Transparency, energy performance and variable geometry: Of the three types of translucent and transparent insulation now on the market (plastic fibres, gas and aerogel), we found that the best core material for the new F²TE³ system is aerogel, (Ismail, 2008; Wong et al., 2007) as it offers:

Transparency: aerogel compares with glass in transparency (87.6%)

Insulation: aerogel is an excellent insulator. According to published data, [4] the thermal performance of a 70 mm nanogel-filled VIP panel is better than a 270 mm cavity filled wall.

Core material: thanks to its porosity and friction joint sealing, silica smoke is the material that many studies (Musgrave, 2009) recommend as VIP core material, provided that there is an outer protective skin. This also has a barrier function in order to maintain the vacuum of the panel.

Lightness: aerogel is only three times heavier than air, and although the density of the aerogel that we use as insulation is higher (50-150kg/m³) remains very light.

Versatility: monolithic aerogel can be shaped as required (Tenpierik & Cauberg, 2006).

5. Conclusion

F^2TE^3 is a slim façade system that provides high energy efficiency, with a seamless surface, providing for variable geometry and the option of building self-supporting structures into the same transparent system skin. The study conducted as part of this research has shown that the prototype F^2TE^3 system outperforms other systems existing on the market, offering added value in terms of structure, transparency and variable geometry, and overcoming the traditional weaknesses of most of the VIP panels.

Computer-simulated trials have shown it to have almost identical energy efficiency properties to monolithic aerogel systems and VIP envelopes. This system revolutionizes VIP systems, as it generates a transparent envelope but eliminates breakages due to fragility by substituting glass for a reinforced composite material. Additionally, it offers the option of generating variable geometry designs.

The prototype F^2TE^3 system outperforms the systems existing on the market by combining some of the best properties of these systems and solving their weaknesses.

6. Acknowledgment

The development of this research work has required the participation of numerous comercial companies, that have provided information and material that made this research feasible. To this end, I quote Wim Van Eynde Makrolon Ambient's Commercial Director (Belgium), Christina Beck, Director of the Office Sales and Export of Okalux (Germany) and Silvestre Shana Market Development Manager of the Cabot Corp. company (USA).

We also want to thank for their collaboration to Anna Roig researcher at the Materials Science Institute of Barcelona (ICMAB-CSIC) and the support of the ABIO research group of E.T.S.A.M., the Department of Construction and Technology in Architecture, School of Architecture (UPM) and the Languages and Systems and Software Engineering Department, School of Computing (UPM).

7. References

Baetens, R., Jelle, B. P., Thueb, J. V., Tenpierikd, M. J., Grynninga, S., Uvsløkka, S. & Gustavsene, A. (2010). Name of paper. *Vacuum insulation panels for building applications: A review and beyond,* Vol. 42, No. issue 2, (February 2010), pp. (147-172), ISSN 0378-7788

Commission of the European Communities. (23.1.2008). *Moving Forward Together on Energy Efficiency* (Communication from the Commission to the Council and the European Parliament on a first assessment of National Energy Efficiency), Action Plans as required by Directive 2006/32/ec on energy end-use efficiency and energy services, 2008, COM(2008) 11 final, Brussels

Fricke, J. (2005). Title of conference paper, *Proceedings of 7th International Vacuum Insulation Symposium,* ISBN 3-528-14884-5, Empa, Duebendorf/Zurich, Switzerland, September 2005

Gibson, A. D., (Aug 2009). *Construction Panel System and Method of Manufacture Thereof,* (PCT), The International Patent System, patno: wo09105468, USA

Heinemann, U., Weinläder, H. & Ebert, H.-P., (03 March 2009). *Energy efficient building envelopes: New materials and components* (Energy Working Group (EPA)), German Research Centre in terms of energy applications, IBS/b - 43145/00, Hamburg

Ismail, K.A.R. (2008). Comparison between PCM filled glass windows and absorbing gas filled windows. *Energy and Buildings*, Vol. 40, No. 5, (July 2008), pp. (710-719), ISSN: 03787788

Jensen, K. I., Kristiansen, F. & Schultz, J. (Eds.). (30 November 2005). *HILIT+. Highly Insulating and Light Transmitting Aerogel Glazing for Super Insulating Windows*, Department of civil engineering (BYG-DTU) Technical University of Denmark, Contract ENK6-CT-2002-00648, Lyngby, Denmark

Martín, L., Oriol-Ossó, J., Ricart, S., Roig, A., García, O. & Sastre, R., (2008). : Organo-modified silica Aerogels and implications for material hydrophobicity and mechanical properties. *Journal of Materials Chemistry*, No. 18, (month and year of the edition), pp. (207 - 213), ISSN 0959-9428

Moner-Girona, M., Martínez, E. & Roig, A., (2002). Micromechanical properties of carbon–silica Aerogel compositesAerogeles. *Appl. Phys.*, Vol. A, No. 74, (2002), pp. (119–122), ISSN 0947-8396

Mukhopadhy, P., Kumaran, K., Ping, F. & Normandin, N. (2011). Use of Vacuum Insulation Panel in Building Envelope Construction: Advantages and Challenges, *Proceedings of 13th Canadian Conference on Building Science and Technology*, N R C C - 5 3 9 4, Winnipeg, Manitoba, May, 2011

Musgrave, D. (2009). Structural Vacuum Insulation panels, *Proceedings of International Vacuum Insulation Symposium (IVIS-09)*, ISBN, London, Uk, September 2009

Rubin, M. & Lampert M., (1982). Transparent Silica Aerogel for Window Insulation. *Solar Energy Materials*, No. 7, (September 1982), pp. (393-400), ISSN 0165-1633 Irregular

Simmler, H., Brunner, S., (EMPA) Heinemann, U., Schwab, H., (ZAE-Bayern) Kumaran, K., Mukhopadhyaya, P., (NRC-IRC) Quénard, D., Sallée, H., (CSTB) Noller, K., Kücükpinar-Niarchos, E., Stramm, C., (Fraunhofer IVV) Tenpierik, M., Cauberg, H., (TU Delft) & Erb, M., (Dr.Eicher+Pauli AG) (Sept. 2005). *Vacuum Insulation Panels - Study on VIP-components and Panels for Service Life Prediction of VIP in Building Applications (Subtask A-B)*, (HiPTI-High Performance Thermal Insulation), IEA/ECBCS, Annex 39, Switzerland

Tenpierik, M. & Cauberg, H. (2006). Vacuum Insulation Panel: friend or foe?, *Proceedings of PLEA2006, The 23rd Conference on Passive and Low Energy Architecture*, ISBN 2-940156-31-X, Geneva, Switzerland, September 2006

Willems, W. M. & Schild, K. (2008). Where to use vacuum insulation… and where not!, *Proceedings of 8th Nordic Symposium on Building Physics*, ISBN/ISSN: 978-87-7877-265-7, Copenhagen, Denmark, June 2008

Willems, W. M. & Schild, K. (2006). The use of Vacuum Insulated Sandwiches (VIS) in building constructions, *Proceedings European Energy Performance of Buildings Directive at EPIC 2006 AIVC Conference*, ISBN:2-86834-122-5, Lion, France, 20-22 November 2006

Wong, I.L., Eames, P.C., & Perera, R.S., (2007). A review of transparent insulation systems and the evaluation of payback period for building applications. *Solar Energy*, Vol. 81, No. 9, (Sep 2007), pp. (1058-1071), ISSN 0038-092X

5

A New Method for Numerical Modeling of Heat Transfer in Thermal Insulations Products

Sohrab Veiseh
Building and Housing, Research Center
Iran

1. Introduction

Mineral fiber products are the most common group of thermal insulations currently in use. Heat transfer in these materials has been the subject of extensive investigations thanks to their numerous and varied applications as building and industrial insulations. Early studies addressed the problem of energy transport in terms of a simple theoretical model. They showed that gas conduction and radiation are the two dominant modes of heat transfer in fibrous insulations [Verschoor et.al. (1952), Bankvall (1974), Bhattacharyya (1980), and Larkin and Churchill (1959)].

Subsequent theoretical studies have been devoted to the solution of the radiative transfer equation in semi-transparent absorbing and isotropic scattering media. Tong and Tien (1980) developed analytical models for radiation in fibrous insulations. They (1983) modeled the radiative heat transfer by the two-flux and linear anisotropic scattering solutions compared well with experimental values. Transient heat transfer was also studied in other works Tong et. al. (1985-1986), and McElroy (1986).

Lee (1986, 1988, and 1989) and Lee and Cunnington (1998, and 2000) proposed radiation models which rigorously account for fiber morphology and orientation. Later models (1997, and 1998) used the radiative properties of the fibers. The contribution of radiative heat transfer through foam insulation was examined by Glicksman et al. (1987). Langlais et al. (1995) worked with the spectral two-flux model to analyze the effect of different parameters on radiative heat transfer. Zeng et al. (1995) developed approximate formulation for coupled conduction and radiation through a medium with arbitrary optical thickness. Daryabeigi (1999) developed an analytical model for heat transfer through high-temperature fibrous insulation. The optically thick approximation was used to simulate radiation heat transfer. He (2003) also modeled radiation heat transfer using the modified two-flux approximation assuming anisotropic scattering and gray medium.

Asllanaj, Milandel and their coworkers (2001, 2002, 2004, and 2007) studied different aspects of radiative-conductive heat transfer in fibrous media and made great contribution to the progress of this field. Yuen et al. (2003) used measured optical properties, the Mie theory, and the zonal method, to predict the transient temperature behavior of fibrous insulation.

Nisipeanu and Jones (2003) applied the Monte Carlo method to model radiation in the entire coarse fibrous media. Not only is this method computationally demanding, it also fails to take into account the contribution of conduction in radiation heat transfer. Moreover, it assumes random distribution of fibers in the media, while in reality the majority of fibers are oriented perpendicular to the heat flux.

The Monte Carlo method is essentially a time consuming process. As such, it has not been widely applied to model radiation heat transfer in previous studies. In the present work, however, distribution factors have been used to expedite computation. The number of calculations during each iteration is considerably reduced by this method. Radiation is coupled with conduction via the source term in the heat conduction equation. In addition, the present method considers fiber orientation perpendicular to heat flux, which is a more logical assumption than random orientation of fibers.

2. Physical model and mathematical formulation

As depicted in Fig. 1, the analytical model assumes that insulation is confined between two horizontal plates, having temperatures T_H (top plate) and T_C (bottom plate). Thus, the heat flux vector is aligned with the local gravity vector in order to eliminate free convection. Air inside the material is considered to be stagnant and dry and at atmospheric pressure. The heat transfer mechanism in fibrous insulations therefore includes solid conduction, gas conduction and radiation and the total heat flux is given by the sum of radiative and conductive heat fluxes:

$$q_t = q_c + q_r \tag{1}$$

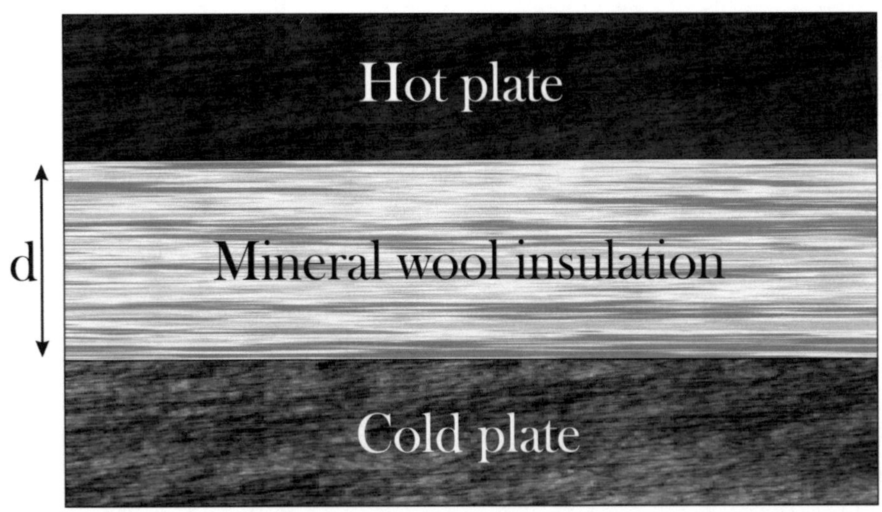

Fig. 1. Problem geometry

The steady state energy equation for a one-dimensional heat transfer is given by:

$$\frac{d}{dx}\left(k_c(T)\frac{dT}{dx}\right) - \frac{dq_r(x)}{dx} = 0 \tag{2}$$

where $k_c(T)$ is the effective thermal conductivity of the medium. The semi-empirical relation suggested by Langlais and Klarsfeld (2004) was used to model $k_c(T)$ for insulations made of silica fibers. The relation is based on experimental data obtained from a Guarded Hot Plate apparatus (Saint Gobain Research Center):

$$k_c(T) = 10^{-3}\left(0.2572\,T_m^{0.81} + 0.0527\,\rho^{0.91}\left(1 + 0.0013\,T_m\right)\right) \tag{3}$$

where ρ is the bulk density of the fibrous insulation, and T_m is the mean temperature of the medium. This relation takes into account both air and fiber conduction as well as the contacts between fibers.

3. Radiation modeling

Radiation heat transfer through the medium considered in this work involves absorption, emission and scattering. The radiation modeling introduced here is based on the Monte Carlo Ray Trace (*MCRT*) method [Modest (2003), and Mahan (2002)], a statistical approach in which analytical solution of the problem is bypassed in favor of a numerical simulation, which is easier to carry out. The probabilistic description of radiation heat transfer by the *MCRT* method [Modest (2003), and Mahan (2002)] is based on the photon view of thermal radiation. The general approach in the *MCRT* method is to emit a large number of energy bundles from randomly selected locations on a given surface element and then to trace their progress through a series of reflections until they are finally absorbed on a surface element [Mahan (2002)]. As radiation heat transfer is a three dimensional phenomenon, direct simulation is utilized to model radiation heat transfer in fibrous media.

Equation (2) is a one-dimensional energy equation therefore it should be coupled with one dimensional radiative heat transfer equation. Accordingly, results for a three-dimensional direct modeling need to be averaged out into a one-dimensional media. The radiative heat flux term in Eq. (2) indicates a radiative heat source. Therefore radiative heat sources have to be found in parallel planes along the x-axis. The one-dimensional radiation heat transfer equation for computing these radiative heat sources can be written as:

$$\begin{aligned} q_r(x) &= \sigma\left(\varepsilon_H D_{Hi} T_H^4 - \varepsilon_i D_{iH} T_i^4\right) + \sigma\left(\varepsilon_i D_{iC} T_i^4 - \varepsilon_C D_{Ci} T_C^4\right) \\ &+ \sigma \sum_{j=1}^{i}\left(\varepsilon_j D_{ji} T_j^4 - \varepsilon_i D_{ij} T_i^4\right) + \sigma \sum_{j=i+1}^{n}\left(\varepsilon_i D_{ij} T_i^4 - \varepsilon_j D_{ji} T_j^4\right) \end{aligned} \tag{4}$$

where i indicates the plane number at the location of x, H, and C indicate the hot and cold bounding plates. D_{Hi} and D_{Ci} are the the radiation distribution factor (*RDF*) of the hot and cold bounding plates to the fibers' plane respectively, D_{iH} and D_{iC} are the *RDF* of the fibers' plane to the hot and cold bounding plates respectively. D_{ij} and D_{ji} are the *RDF* between different elements within the media. D_{ij} is defined as the fraction of the total radiation emitted diffusely from element i and absorbed by element j, due to both direct radiation and to all possible diffuse and specular reflections within the enclosure [Mahan (2002)].

The first term on the right hand side of Eq. (4) represents the radiative heat flux emitted by the hot bounding plate and absorbed by the fibers in plane *i*, the second term represents the radiative heat flux emitted by the fibers in plane *i* and absorbed by the cold bounding plate, and the third and fourth terms represent the resultant interaction radiative heat flux between the fibers in plane *i* and other fibers within the media in different planes.

From Eq. (4), it is clear that the radiative distribution factor is required for two different cases; *RDF* among fibrous planes and the *RDF* of the fibrous planes to the boundary plates. Hence, the problem is to find the radiation distribution factor for these two cases as a function of relative distance.

Application of the reciprocity relation, Eq. (5), readily gives the distribution factors from other planes to the source plane. A similar procedure is adopted to compute the *RDF* of the fibers to the bounding plates.

$$\varepsilon_i A_i D_{ij} = \varepsilon_j A_j D_{ji} \tag{5}$$

where ε_i and ε_j are the emissivity of the fibers *i* and *j*. A_i and A_j are the surface areas of fibers planes i and j.

It is assumed that for the limited temperature range considered, the radiative distribution factor is not a function of temperature. Therefore, it is possible to compute the *RDF* of the fibers for the mean temperature properties and it is not required to recompute distribution factors in each iteration procedure.

In addition, as the fibers are distributed randomly in the plates normal to the heat flux, the *RDF* of the fibers is not a function of their position but of their relative distance. For instance it is possible to say that D_{ij} for fiber *i* has the same value for all fibers *j* which are located at the same distance from fiber *i*. Therefore, it is only required to compute one fiber's *RDF* in the assumed simulated cylindrical media and the results can be utilized for the entire domain. To compute the *RDF* of the fibers to the plates, it is possible to compute the *RDF* of the plate to the fibers and apply the reciprocity rule (Eq. 5).

The following procedure is adopted for computing the *RDF* of the fibers:

A simulated cylindrical media with a specific radius and infinite height is assumed in which fibers are randomly located parallel to cylinder's axis as shown in Fig. 2. It is assumed that the fibers are distributed randomly with a uniform distribution in the media and the number of fibers per volume in the media is a function of the material's porosity. As the average fiber diameter and the porosity of the material are measurable, it is possible to define the number of the fibers in the defined cylinder. The radius of the assumed cylinder should be long enough so that no emitted energy bundle can escape the media. This length is directly related to the optical thickness of the fibers.

Figure 3 shows the flow chart of the *MCRT* for the given problem. *RDF* of the fibers has a rapidly decaying exponential behavior. Hence the cylinder defined for the determination of the RDF could have a short diameter as compared to the thickness of the real fibrous media. This considerably reduces simulation time.

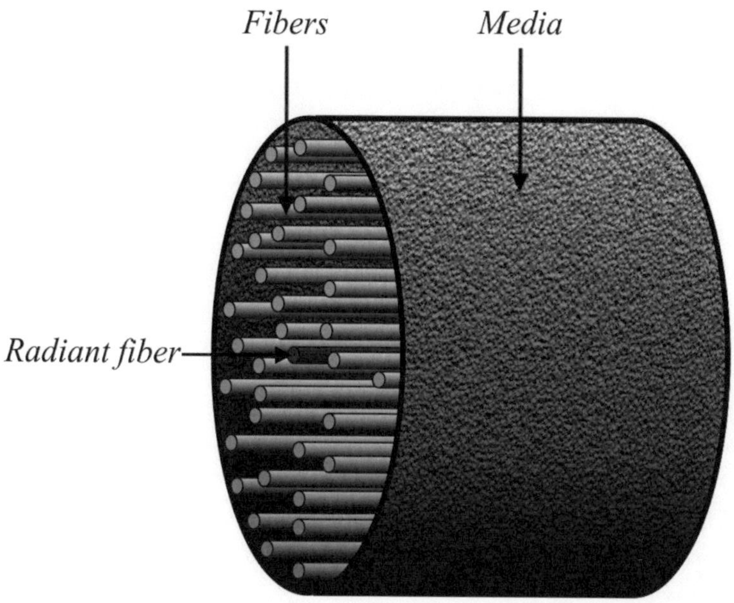

Fig. 2. Simulated cylindrical fibrous media for computing the radiation distribution factor of the fibers

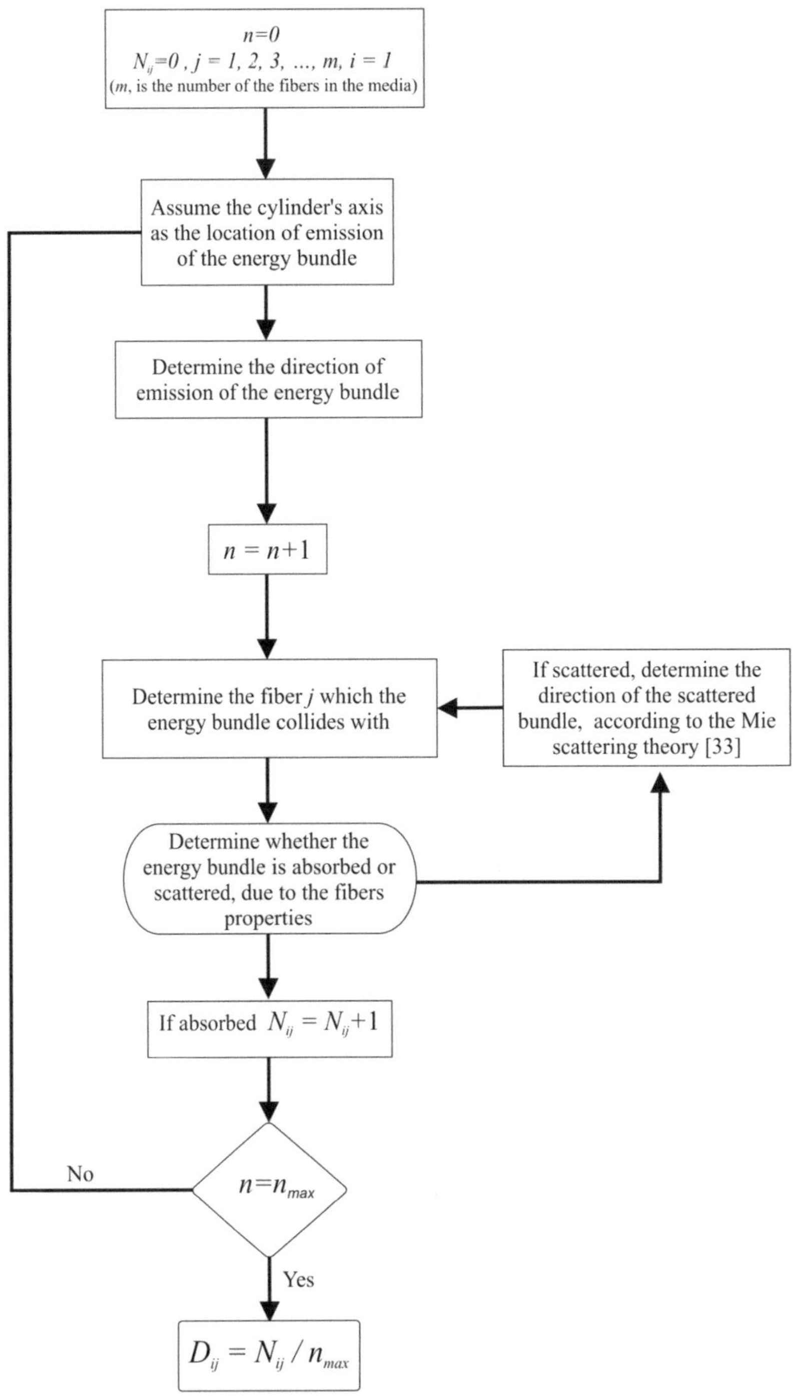

Fig. 3. Flow chart of the MCRT method for computing the fibers radiation distribution factor

The same procedure can be used to compute the distribution factor of the fibers to the bounding plates. In this case the radiant fiber is assumed to be at the center of a semi cylinder, on the boundary surfaces, as shown in Fig. 4. The plate is assumed to be opaque. The *RDF* of the plate to the fibers is determined by direct Monte Carlo simulation, and by employing the reciprocity rule the *RDF* of the fibers to the plates can readily be computed. Figure 5 shows the flow chart of the *MCRT* for the given problem.

The *Mie* scattering phase function is applied to determine the direction of the scattered radiant from fibers [35]. The *Mie* phase function depends on the mean diameter, index of refraction of the fibers and the prominent wave length of the media.

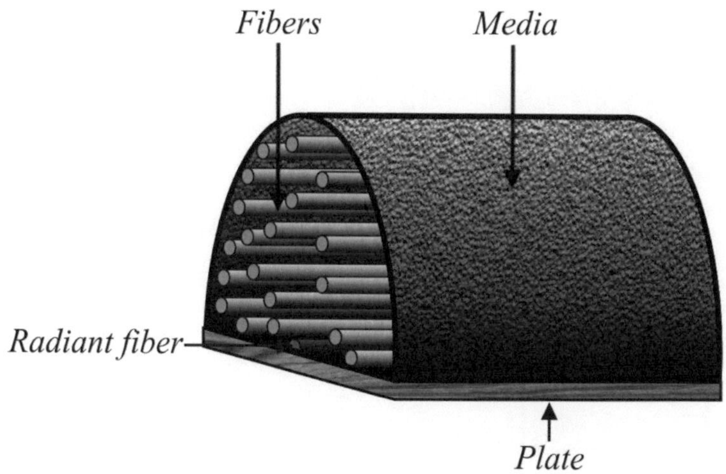

Fig. 4. Simulated semi cylindrical fibrous media for computing the radiation distribution factors of the fibers to the plate

4. Computational procedure

Considering the nature of the problem which involves combined radiation and conduction equations; the solution of the coupled equations involves an iterative procedure. Therefore, in every iterations the conduction and radiation equations should be solved. Since *RDF*s need not be recomputed in every iteration, the computations are considerably more efficient as compared to those methods in which radiation is fully coupled (such as: discrete ordinate method, spherical harmonics, or zonal method).

To solve the energy equation, the simple implicit (Laasonen) method [Anderson (1984)] is used to discretize implicit time and space derivatives. This method has a first-order accuracy with a truncation error of $O[\Delta\tau, (\Delta x)^2]$ and is unconditionally stable. Several grids were tried with 500, 1000, 2500, 5000, and 10000 nodes; comparing the results obtained showed that the 5000 node grid was sufficient for this case study.

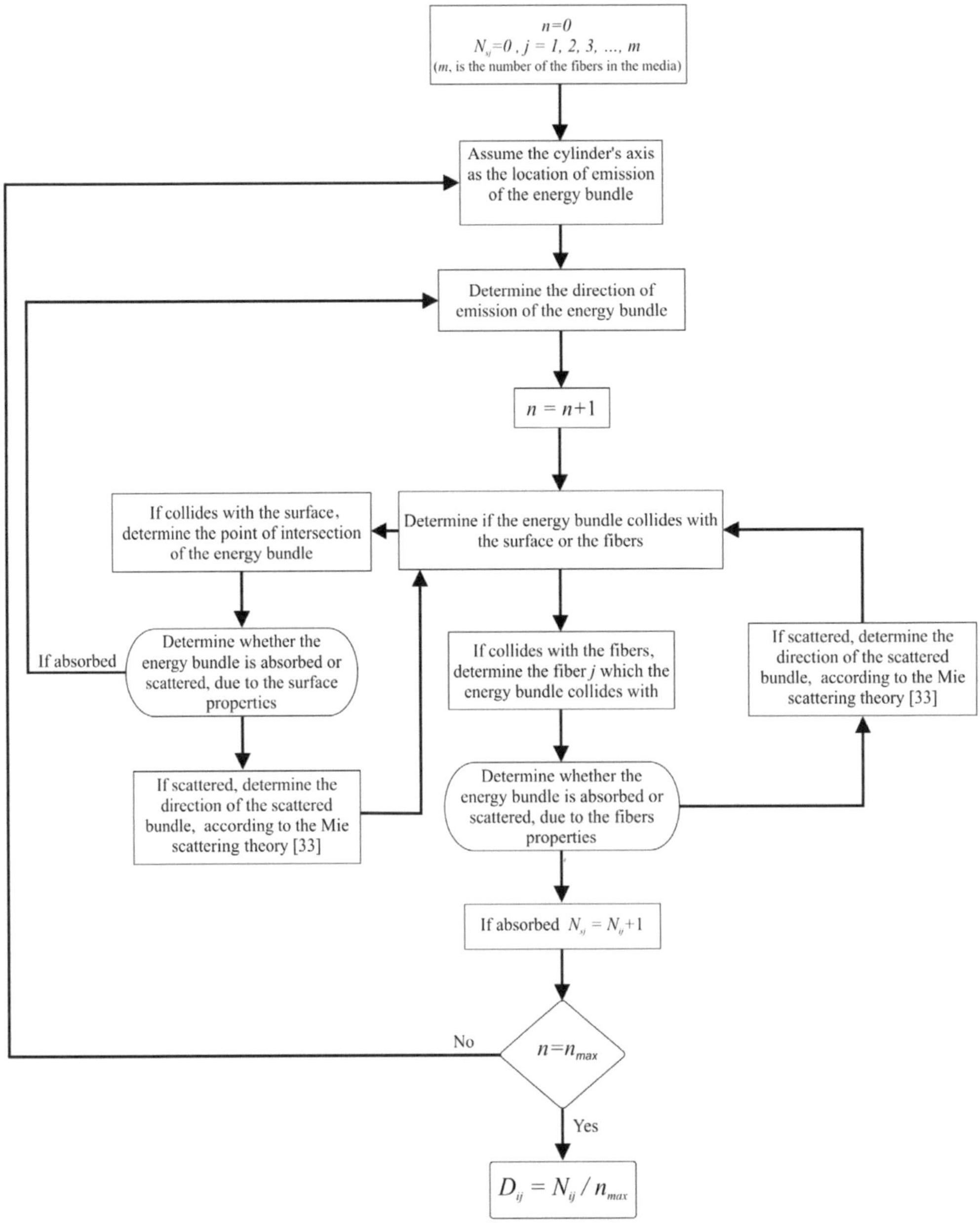

Fig. 5. Flow chart of the MCRT method for computing the fibers radiation distribution factor to the plate

The flow chart of the corresponding method is given in Fig. 6. The iterations continue until convergence of the iterative procedure. The convergence criterion is based on the *rms* of the difference between temperatures of two subsequent iterations as defined below:

$$\sqrt{\frac{1}{n}\sum_{j=1}^{n}\left(T_j^{P+1}-T_j^{P}\right)^2} < 10^{-4} \tag{6}$$

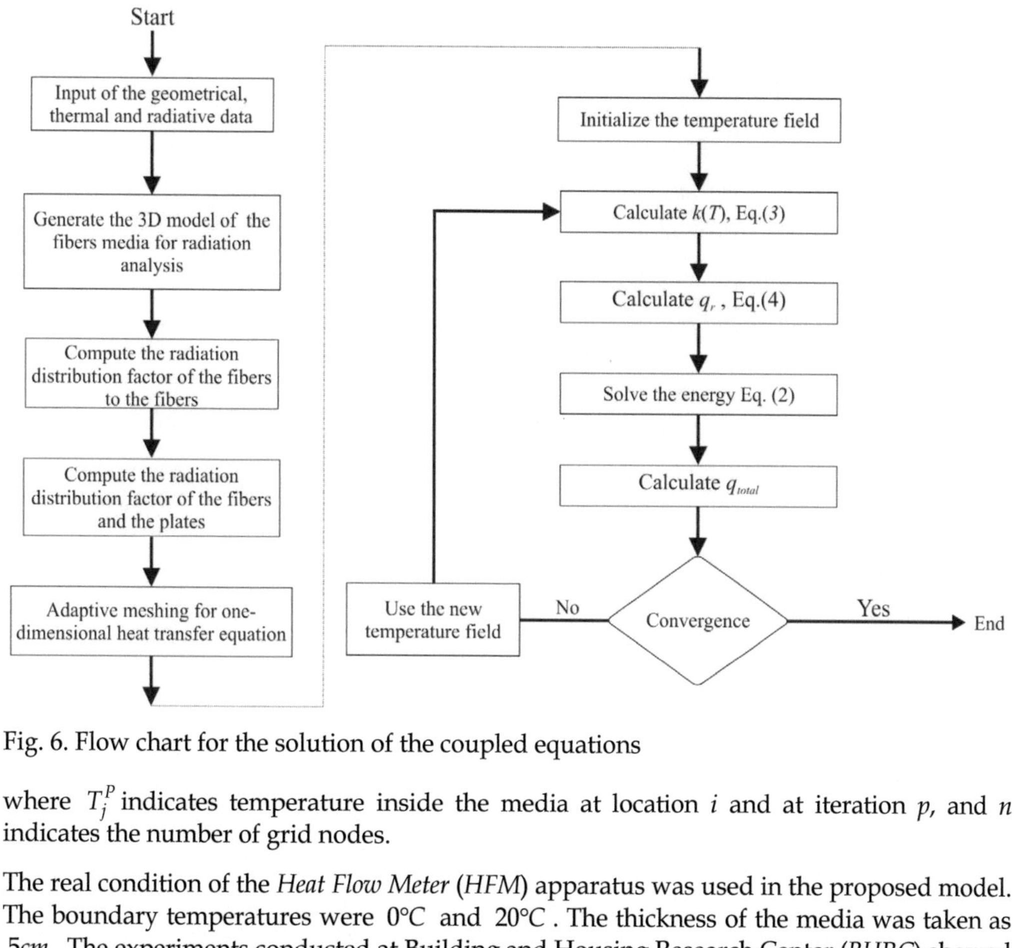

Fig. 6. Flow chart for the solution of the coupled equations

where T_j^P indicates temperature inside the media at location *i* and at iteration *p*, and *n* indicates the number of grid nodes.

The real condition of the *Heat Flow Meter* (*HFM*) apparatus was used in the proposed model. The boundary temperatures were $0°C$ and $20°C$. The thickness of the media was taken as $5cm$. The experiments conducted at Building and Housing Research Center (*BHRC*) showed that the mean diameter of fibers from samples studied was seven microns. The averaged index of refraction for glass is considered to be $1.49+1\times10^{-4}i$, where *i* is imaginary unit (derived from the Hsieh and Su, [Hsieh (1979)]. As the mean temperature is 283K, from Wien's displacement law [Siegel, and Howell (2002)], the wavelength from which the largest amount of radiative energy is transmitted is $\lambda = 10\mu m$. Therefore, the radiative properties are the same as the properties proposed by Roux [Roux (2003)] in this wavelength. A boundary surface emissivity of 0.9 (as declared by Netzsch, the manufacturer of the *HFM* apparatus) is used for these computations.

5. Discussion

5.1 Numerical results

Figure 7(a) shows the cross section of the simulated fibrous media for a density of 500 (kg/m^3) and Fig. 7(b) shows the cross section contour of the radiation distribution

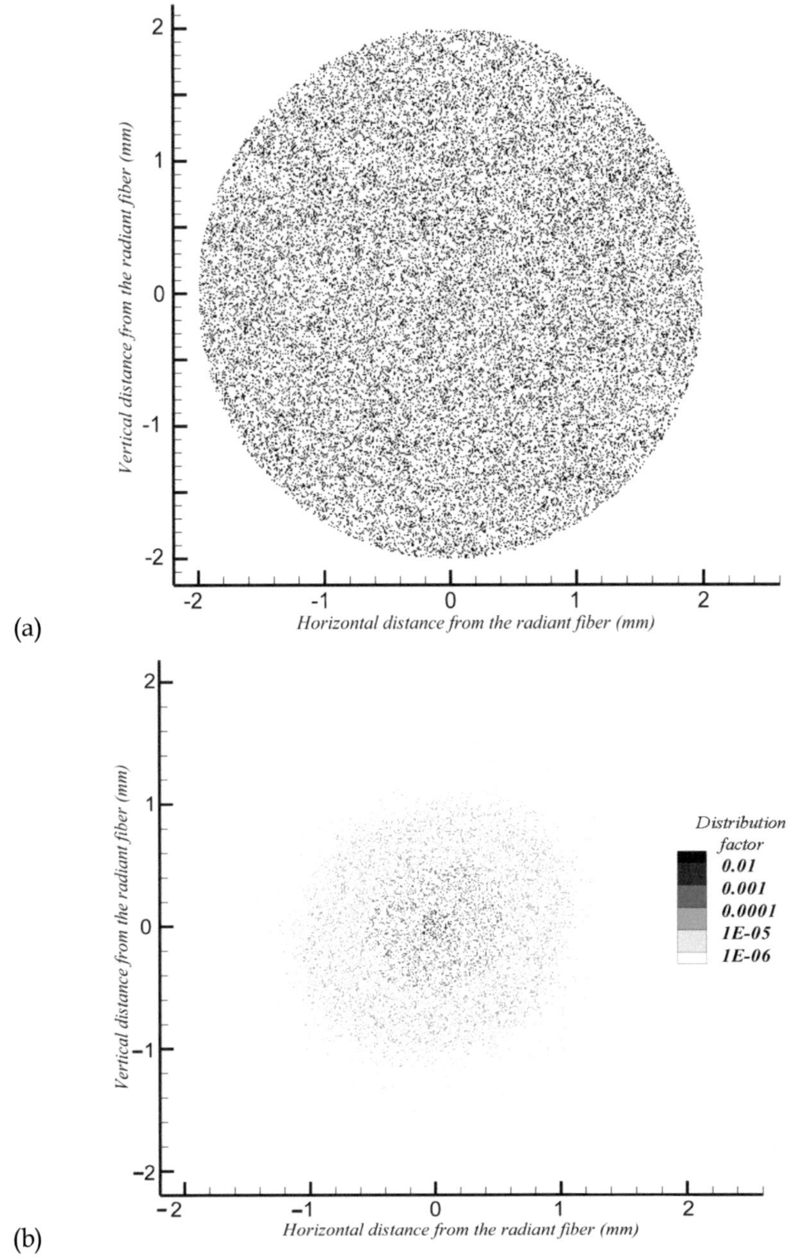

Fig. 7. (a) Cross section of the simulated cylindrical fibrous media for $\rho = 500kg/m^3$, (b) Cross section contour of the radiation distribution factor for $\rho = 500kg/m^3$

factor for the same density. Figure 7 clearly shows that the radiative distribution factor decays rapidly, obviating the need for determination of the *RDF* for the entire media.

The results of the effective thermal conductivity, k_e, (Eq. (7)), radiative conductivity, k_r, (Eq. (8)), and the air and glass fiber conductivity, k_c, (Eq. (9)) computed with the current method for different densities between 5 and 500 (kg / m^3) are shown in Fig. 8.

$$k_e = q_t / (T_H - T_C) \tag{7}$$

$$k_r = q_r / (T_H - T_C) \tag{8}$$

$$k_c = q_c / (T_H - T_C) \tag{9}$$

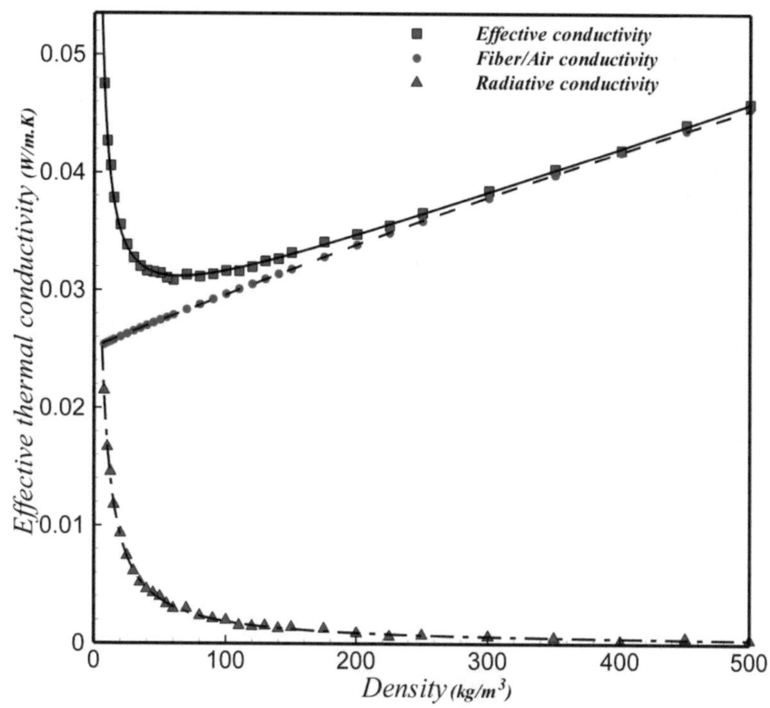

Fig. 8. Effective thermal conductivity, air/fiber conductivity and the radiation conductivity of glass fiber for different densities and mean temperature of 10°C

Total heat flux, conduction and radiation heat flux of fiber glass under steady state condition for $\rho = 50 kg / m^3$ and $\rho = 7.5 kg / m^3$ according to the position in the medium, are shown in Figs. 9(a) and 9(b), respectively. As is seen in Fig. 9(a, b) total heat flux for $\rho = 7.5 kg / m^3$ is 33.8% greater than the total heat flux for the $\rho = 50 kg / m^3$. In addition, the radiation heat flux is 12.6% of the total heat flux for $\rho = 50 kg / m^3$, and 45.2% of the total heat flux for $\rho = 7.5 kg / m^3$.

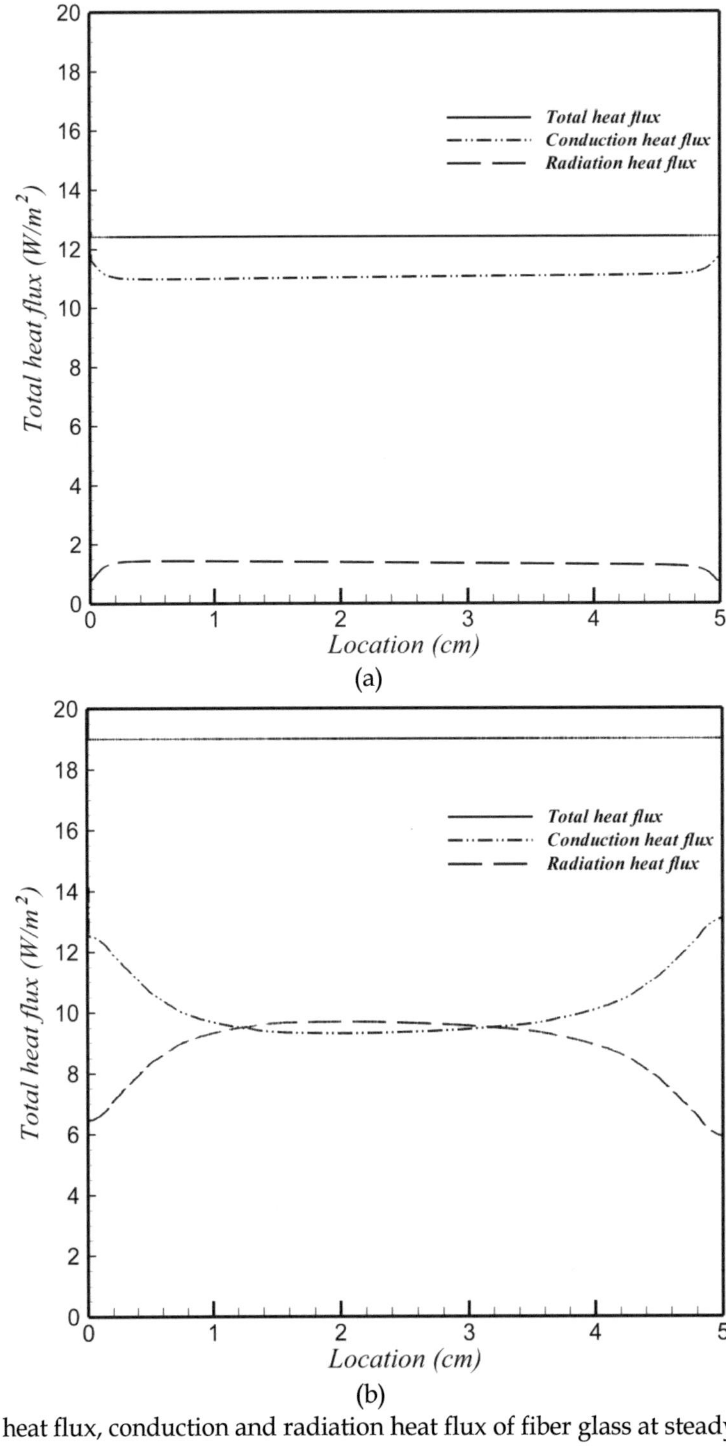

Fig. 9. Total heat flux, conduction and radiation heat flux of fiber glass at steady state condition and mean temperature of 10°C, according to the position in the medium: (a) $\rho = 50kg/m^3$, (b) $\rho = 7.5kg/m^3$

The temperature profiles within the medium for mean temperature of 10°C, and $\rho = 50kg / m^3$ and $\rho = 7.5kg / m^3$ are shown in Fig. 10.

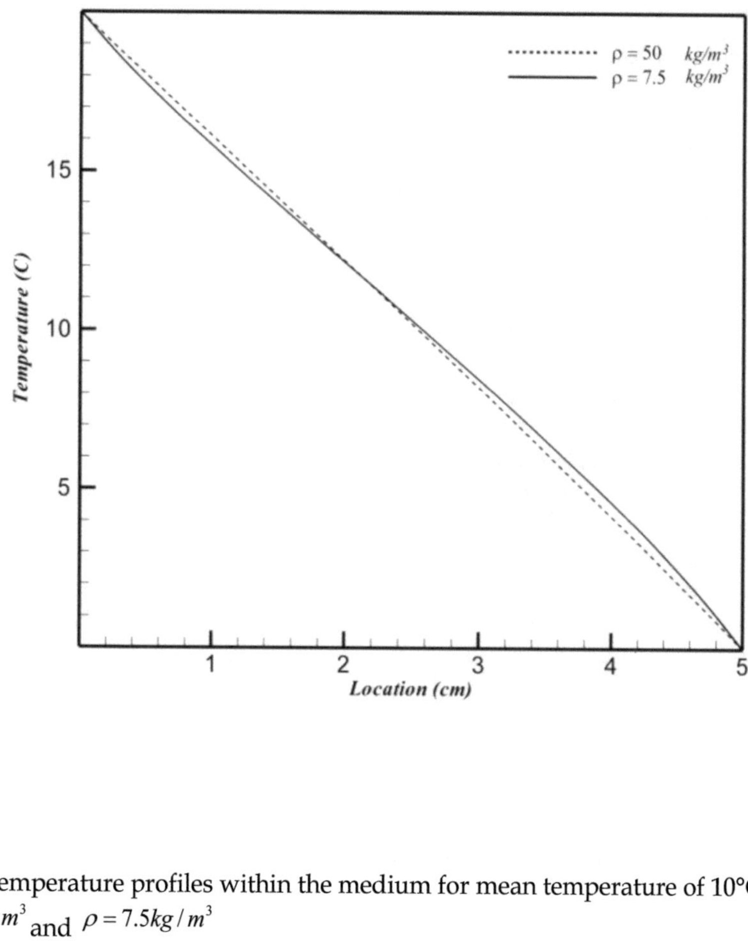

Fig. 10. Temperature profiles within the medium for mean temperature of 10°C, and $\rho = 50kg / m^3$ and $\rho = 7.5kg / m^3$

5.2 Experimental measurements

A large number of experiments were performed at *BHRC* on mineral wool insulations for determination of thermal conductivity and microstructural analysis. The stereo-microscopy observations of the samples showed that most of the fibers are oriented parallel to the faces and the boundaries (Fig. 11). Thus, the direction of heat flow is perpendicular to the direction of the majority of fibers. Accordingly the model's assumption of parallel cylindrical fibers is well justified.

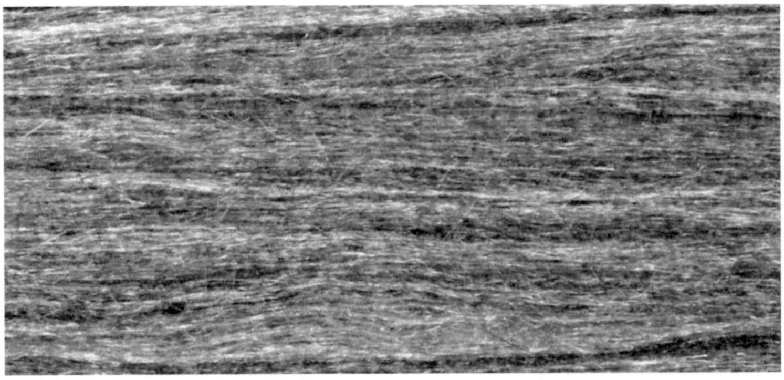

Fig. 11. Stereo micrograph of glass wool

Scanning electron microscopy (*SEM*) showed that the diameter of the fibers is *6.2-8.8* microns. The mean diameter of fibers determined with *SEM* is about $7\ \mu m$. As the diameters of these fibers are much smaller than their length, the length can be assumed infinite in the model in comparison to the diameter.

The effective thermal conductivity of more than 300 different samples of glass fiber products with densities ranging from 6 to $120\ kg/m^3$ were measured. The conductivity measurements were carried out at *BHRC* with a heat flow meter (*HFM*) apparatus according to *EN12667* (2001). The *HFM* apparatus used is a single-specimen symmetrical device that consists of two heat flow meters and allows the detection of the heat transfer rate on both the hot and cold sides of the specimen. The cold and hot plate temperatures were set at $0°C$ and $20°C$, respectively. The samples were dried in a ventilated oven and then brought into equilibrium with laboratory air temperature. To prevent moisture from migrating to the specimens during the test, specimens were enclosed in a vapor-tight envelope. The accuracy of thermal conductivity determination was better than ±3%. Measurement repeatability was found to be better than ±1% both when the specimen was maintained in the apparatus and removed and mounted after a long interval. For bulk density determination the accuracy in the measurements of specimen length, width, and thickness were better than ±1%. The maximum uncertainty in measured specimen thickness due to departures from a plane was 0.5%. The maximum uncertainty in the determination of specimen mass was 0.5%.

5.3 Comparison of numerical and experimental results

The numerical model was validated by comparing the predicted effective thermal conductivity with measured data from the research at *BHRC*, those obtained in Technical Research Institute of Sweden (*SP*) [Jonsson (1996)], and those presented in ASHRAE handbook [ASHRAE (1997)] for fibers with $5.6\ \mu m$ diameter.

The comparison of the effective thermal conductivity of glass fiber having different densities obtained from the proposed model and the experimental results, are shown in Fig. 12. It can be seen that in lower densities, where radiation is dominant, experimental results conform excellently to the model predictions. Table 1 shows the percentage of difference between the results of the proposed model and the experimental results. The model predictions are in good agreement with measurement results.

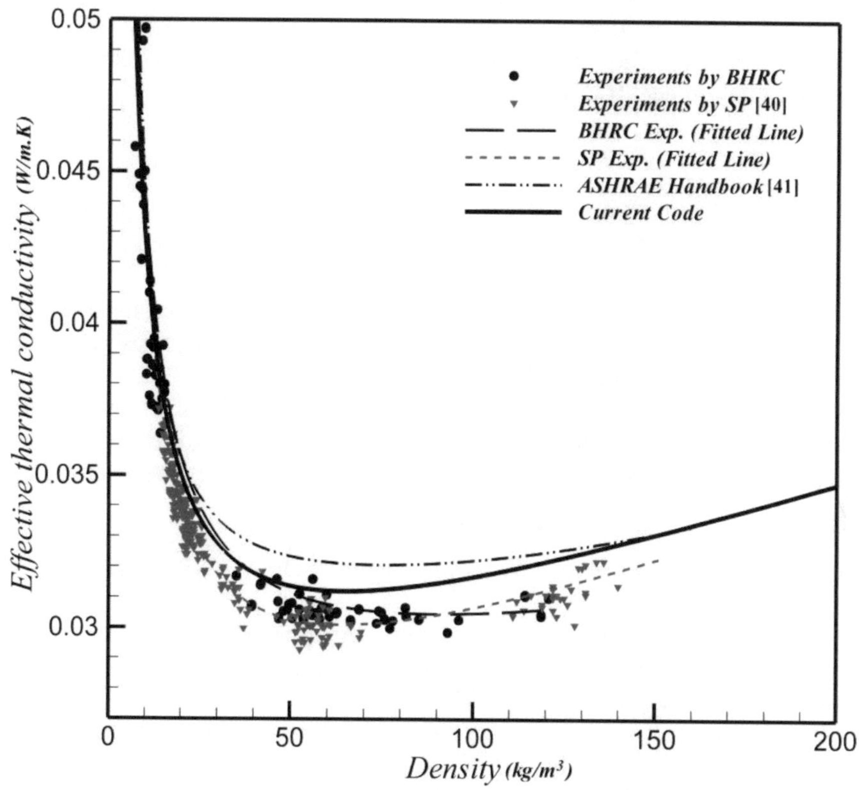

Fig. 12. Comparison of effective thermal conductivity between current method and experimental results of SP [Jonsson (1996)] and ASHRAE [ASHRAE (1997)] and those obtained in a research project at BHRC

Experiments done by	Density Range (kg/m^3)	Percent difference of proposed model and experiments
SP [40]	**10-140**	**3.3 %**
ASHRAE [41]	**8-160**	**2.6 %**
BHRC	**6-120**	**1.3 %**

Table 1. Percent difference between effective thermal conductivity of the model and the experiments by SP, ASHRAE and performed in a research project at BHRC

6. Conclusion

This chapter introduces a new numerical modeling of steady state heat transfer for combined radiation and conduction in a fibrous medium for the prediction of the effective thermal conductivity. Radiant heat transfer in mineral wool insulations is modeled by the statistical-based Monte Carlo method. A finite difference approach has been developed to solve the governing coupled radiation and conduction heat transfer equations. The numerical model was validated by comparison with effective thermal conductivity measurements at different densities. The proposed method is easy to code and computationally efficient. The model is able to sort out individual contributions of conduction and radiation heat transfer mechanisms in these materials.

7. References

Abu-Eishah S. I., Haddad Y., Solieman A., and Bajbouj A., 2004, "A New Correlation for the Specific Heat of Metals, Metal Oxides and Metal Fluorides as a Function of Temperature", Bahia Blanca, Vol. 34, No. 4, pp 257-265.

Andersen, F. M. B., and Dyrbol, S. , 1997, "Comparison of Radiative Heat Transfer Models in Mineral Wool at Room Temperature", *Proc. 2nd International Symposium on Radiation Transfer,* Kusadasi, Turkey, vol. 1, pp. 607-619.

Andersen F. M. B., and Dyrbol, S. , 1998, "Modeling Radiative Heat Transfer in Fibrous Material: The Use of Plank Properties Compared to Spectral and Flux-Weighted Properties", *Journal of Quantitative Spectroscopy and Radiative Heat Transfer,* pp. 593-603.

Aronson J. R., Emslie A. G., Ruccia F. E., Smallman C. R., Smith E. M., and Strong P.F., 1979, "Infrared emittance of fibrous materials", Applied Optics, Vol. 18, No. 15, pp.2622-2633.

ASHRAE, 1985, "Design Heat Transmission Coefficient", ASHRAE Handbook, Fundamental, chap. 23, *American Society of Heating, Refrigerating and Air-conditioning Engineers,* Atlanta, GA, pp. 23.1-23.22.

ASHRAE, 1993, "Thermal Insulation and Vapor Retarders-Fundamentals", ASHRAE Fundamentals Handbook (SI), pp. 20.1-20.21.

ASHRAE, 1997, "Heat Flow Factors Affecting Thermal Performance", ASHRAE Fundamentals Handbook (SI), pp. 22.4-22.5.

Asllanaj, F., Brige, X, and Jeandel, G. , 2007, "Transient Combined Radiation and Conduction in a One-dimensional Non-gray Participating Medium with Anisotropic Optical Properties Subjected to Radiative Flux at the Boundaries", *Journal of Quantitative spectroscopy and Radiative Transfer,* vol. 107, pp. 17-29.

Asllanaj F., Lacroix D., Jeandel G., Roche J., April 2003, "Transient Combined Radiation and Conductive Heat Transfer in Thermal Fibrous Insulation", *Proceeding of Eurotherm 73 Computational Thermal Radiation in Participating Media 15-17, Mons, Belgium.*

Asllanaj, F., Jeandel, G., and Roche, J. R. , 2001, "Numerical solution of Radiative Transfer Equation Coupled with Nonlinear Heat Conduction Equation", *International Journal of Numerical Methods for Heat and Fluid Flow,* vol. 11, no. 5, pp. 449-472.

Asllanaj, F., Jeandel, G., Roche, J. R., and Lacroix, D. , 2004, "Transient combined radiation and conduction heat transfer in fibrous media with temperature and flux boundary conditions", *International Journal of Thermal Sciences,* vol. 43, pp. 939–950.

Asllanaj F., Jeandel G., Roche J. R. , 2001, "Numerical solution of Radiative Transfer Equation Coupled with Nonlinear Heat Conduction Equation", *International Journal of Numerical Methods for Heat and Fluid Flow, 11, 5,* pp. 449-473.

Asllanaj, F., Milandri, A., Jeandel, G., and Roche, J. R., 1999, "Transfert de Chaleur Par Conduction et Rayonnement en Regime Transitoire dans Les Milieux Fibreux", IVe

Colloque interuniversitaire franco-québécois Thermique des systèmes à température modérée.

Asllanaj, F., Milandri, A., Jeandel, G., and Roche, J. R. , 2002, "A Finite Difference Solution of Non-linear Systems of Radiative-conductive Heat Transfer Equations", *International Journal of Numerical Methods in Engineering,* vol. 54, pp. 1649-1668.

ASTM C612-93, 2002, "Mineral Fiber Block and Board Insulation, Specification for", *American Society for Testing and Materials,* Philadelphia.

ASTM C1335-96, 2005, Standard Test Method for Measuring Non-Fibrous Content of Man-Made Rock and Slag Mineral Fiber Insulation, American Society for Testing and Materials, Annual Book of ASTM, part 04.06.

Bankvall, C.G. , 1972, "Natural Convective heat transfer in insulated structures", *lund Institute of Technology,* Report 38.

Bankvall C.G. , May 1973, "Heat Transfer in Fibrous Materials", Journal of Testing and Evaluation, Vol. 1, No. 5, pp. 235-243.

Bankvall C. G., 1974, "Mechanisms of Heat Transfer in Permeable Insulation and Their Investigation in a Special Guarded Hot Plate", Heat transmission measurements in thermal insulations, ASTM STP544, American Society for Testing and Materials, pp. 34-88.

Bhattacharyya R. K. , 1980, "Heat Transfer Model for Fibrous Insulations", Thermal Insulation Performance, ASTM STP 718, D. L. McElroy and R. P. Tye, Eds., American Society for Testing and Materials, pp.272-286.

Bommerg M., Klarsfeld S., January 1983, "Semi-empirical Model of Heat Transfer in Dry Mineral Fiber Insulations", *Journal of Thermal Insulation,* Volume 6, pp. 156-173.

Boulet P., Jeandel G., Morlat G., 1993, "Model of Radiative Transfer in Fibrous Media-Matrix Method", *Int. J. Heat Mass Transfer,* 36, pp. 4287-4297.

Boulet P., Jeandel G., Morlat G., Silberstein A., and Dedianous P., 1994, "Study of Radiative Behavior of Several Fiberglass Materials", *Thermal Conductivity* 22, edited by T. W. Tong, Technomatic, Lancaster, PA, pp. 749-759.

Budaiwil I., Abdou A., and Al-Homoud M., 2002, "Variations of Thermal Conductivity of Insulation Materials under Different Operating Temperatures: Impact on Envelope-Induced Cooling Load", Journal of Architectural ENGINEERING, DECEMBER, pp. 125-132.

Buttner D., Fricke J., Reiss H. , 1985, "Analysis of the Radiative and Solid Conduction Components of the Thermal Conductivity of an Evacuated Glass Fiber Insulation: Measurement with a 700×700 mm^2 Variable Load Guarded Hot Plate Devise", in proceedings, 20th AIAA Thermophysics Conference, Williamsburg, Va., pp. 85-1019.

Cabannes F., Maurau J.C., Hyrien M., Klarsfeld S.M., 1979, "Radiative heat transfer in fiberglass insulating materials as related to their optical properties", High Temperatures - High Pressures, Vol.11, pp. 429-434.

Cohen L. D., Haracz R. D., Cohen A., and Acquista C., March 1983, "Scattering of Light from arbitrarily oriented finite cylinders", *Applied Optics,* Vol. 22, No. 5.

Cunnington G. R., and Lee S. C., 1996, "Radiative Properties of Fibrous Insulations: Theory Versus Experiment", J. of Thermophysics and Heat Transfer, Vo.10, No. 3, pp. 460-466.

Daryabeigi K., June 2002, "Heat Transfer in High-Temperature Fibrous Insulation", *8th AIAA/ASME joint Thermophysics and Heat Transfer Conference, 24-26, St.Louis, MO.*

Daryabeigi, K., 1999, "Analysis and Testing of High Temperature Fibrous Insulation for Reusable Launch Vehicles", American Institute of Aeronautics and Astronaumics, AIAA-99-1044.

Degenne, M., Klarsfeld S., Barthe, M-P., 1978, "Measurement of the Thermal Resistance of Thick Low-Density Mineral Fiber Insulation", *Thermal transmission Measurements of Insulations,* ASTM STP 660, R. P. Tye, Ed., American Society for Testing and Materials, pp.130-144.

Dent, R. W. Skelton, J. and Donovan, J. G., 1990, "Radiant Heat Transfer in Extremely low Density Fibrous Assemblies", Insulation Materials, Testing, and Applications, ASTM STP 1030, D.L. McElory and J. F. Kimpflen, Eds., American Society for Testing and Materials, Philadelphia, pp. 79-105.

Dyrbol S., Elmroth A., Oct. 2002, "Experimental Investigation of the Effect of Natural Convection on Heat Transfer in Mineral Wool", Journal of Thermal Envelope & Building Science, Vol. 26 Issue 2, p153, 12p.

Edmunds W. M. , 1989, "Residential Insulation", *Energy conservation and ASTM standards.*

EN 12664:2002, European Standard, Thermal performance of building materials and products - Determination of thermal resistance by means of guarded hot plate and heat flow meter methods - Dry and moist products of medium and low thermal resistance.

EN 12667:2001, European Standard, Thermal performance of building materials and products - Determination of thermal resistance by means of guarded hot plate and heat flow meter methods - Products of high and medium thermal resistance, European Committee for Standardization

EN 12939: 2000, European Standard, Thermal performance of building materials and products - Determination of thermal resistance by means of guarded hot plate and heat flow meter methods - Thick products of high and medium thermal resistance.

EN13162: 2001, European Standard, Thermal insulation products for buildings - Factory made mineral wool (MW) products – Specification.

Endriukaityte A., Bliudžius R.., Samajauskas R., 2004, "Investigation of Hydrothermal Performance of Fibrous Thermal Insulation Materials", Materials Science, Vol. 10, No. 1.

Fournier D., Klarsfeld S., 1974, "Some Recent Experimental Data on Glass Fiber Insulating Materials and Their Use for a Reliable Design of Insulations at Low Temperatures", *Heat transmission Measurements in Thermal Insulations, ASTM STP 544,* American Society for Testing and Materials, pp.223-242.

Frances De Ponte, 1985, "Present and Future Research on Guarded Hot Plates and Heat Flow Meter Apparatus", *American Society for Testing and Materials, Philadelphia, pp. 101-120.*

Fricke J., Buttner D., Caps R., Gross J., Nilsson, O., 1990, "Solid conductivity of Loaded Fibrous Insulations", *Insulation Materials, Testing and applications, ASTM STP 1030,* D.L. McElory and J.F. Kimpflen, Eds., American Society for Testing and Materials, Philadelphia, pp. 66-78.

Fricke J., Caps R., Hummer E., Doll G., Arduini M. C., De Ponte F., 1990, "Optically Thin Fibrous Insulations", Insulation Materials, Testing and Applications, ASTM STP 1030, D. L. McElory and J. F. Kimpflen, Eds., American Society for Testing and Materials, Philadelphia, pp. 575-586.

Guyer E. C., Brownell C. L., 1999 , "Handbook of Applied Thermal Design", Taylor & Francis.

Goo N. S., Woo K., *2003,* "Measurement and Prediction of Effective Thermal Conductivity for Woven Fabric Composites", *International Journal of Modern Physics B, Vol. 17, Nos. 8 & 9, 1808-1813.*

Guilbert G., Langlais C., Jeandel G., Morlot G., Klarsfeld S. , 1987, "Optical Characteristics of Semitransparent Porous Media", *High Temperatures – High Pressures,* Vol. 19, pp. 251-259.

Gustafsson S. E., Karawacki E., 1991, "Thermal Transport in Building Materials", *Swedish Council for Building Research, Stockholm.*

Hager N. E., Steere R. C., November 1967, "Radiant Heat Transfer in Fibrous Thermal Insulation", *Journal of Applied Physics,* vol. 38, No. 12, pp. 4663-4668.

Hokoi, S., Kumaran, M., K. , 1993, "Experimental and analytical investigations of simultaneous heat and moisture transport through glass fiber insulation", *J. of Building Physics,* 16(3): pp. 263-292.

Houston R. L. and Korpela S. A., 1982, "Heat Transfer Through Fiberglass Insulation", Proc. Of the Seventh International Heat Transfer Conference, Munch, 2,pp. 499-504.

Huetz-aubert, M., Klarsfeld, S., 1995, "Rayonnement thermique des matériaux semi-transparents", Bases physiques, *Techniques de l'ingénieur Extrait de la collection,* B 8215, pp.26-27.

ISO 8301:1991 Thermal insulation - Determination of steady state thermal resistance and related properties - Heat flow meter apparatus.

Jauen J. L., Klarsfeld S., 1987, "Heat Transfer Through a Still Air Layer", *Thermal Insulation: Materials and systems,* ASTM STP 922, F. J. Powell and S. L. Mattews, Eds., American Society for Testing and Materials, Philadelphia, pp. 283-294.

Jonsson B., 1996, "The Relationship Between Thermal Conductivity and Density for Mineral Wool and Expanded Polystyrene", *Proceedings of the 4th Symposium of Bilding Physics in the Nordic Countaries,* Finland, pp. 675-682.

Keller K., and Blumberg J., 1990, "High Temperature Airborne Fiber Insulations Heat Transfer", *Proceedings of the Ninth International Heat Transfer Conference,* Vol. 5, edited by G. Hetsroni, Hemisphrer, pp. 479-484.

Kielmeyer W. H. and Troyer R. L. , 1999, "Fibrous Insulations", Handbook of Applied Thermal Design, E.C. Guyer and C.L. Brownell, Eds., Taylor & Francis, pp.3.12-3.22.

Klarsfeld S., Boulant J., and Langlais C., 1987, "Thermal Conductivity of Insulants at High Temperature: Reference Materials and Standards", *Thermal Insulation: Materials and Systems, ASTM STP 922,* F. J. Powell and S. L. Matthews, Eds., American Society for Testing and Materials, Philadelphia, pp. 665-676.

Kumaran M. K. and Stephenson D. G., 1988, "Heat transport through fibrous insulation materials", *J. of Building Physics,* 11(4): pp. 263-269.

Langlais, C. and Boulant, J., 1990, "Use of Two Heat transducers for transient thermal measurements on porous insulating Materials", Insulation Materials, testing and applications, ASTM STP 1030, D. L. McElroy and J. F. Kimpflen, Eds. American Society for Testing and Materials, Philadelphia, PP. 510-521.

Langlais, C., Guilbert, G., and Klarsfeld, C., 1995, "Influence of the Chemical Composition of Glass on Heat Transfer through Glass Fiber Insulations in Relation to Their Morphology and Temperature Use", *Journal of Thermal Insulation and Building Envelopes*, vol. 18, pp. 350-376.

Langlais C., Hyrien M., Klarsfeld S., 1983, "Influence of Moisture on Heat Transfer through Fibrous Insulating Materials", *Thermal Insulation, Materials and systems for Energy Conservation in the 80s*, ASTM STP 789, F. A. Govan, D. M. Greason, J. D. McAllister, Eds., American Society for Testing and Materials, Philadelphia, pp. 563-581.

Langlais C., Klarsfeld S., 1985, "Transfert de chaleur a travers les isolants fibreux en relation avec leur morphologie", *Journée D'etudes sur les Transferts Thermiques dans les Isolants Fibreux*, pp. 1-34.

Langlais, C. Klarsfeld, S., 2004, "Isolation thermique à température ambiante", Bases physiques, *Techniques de l'ingénieur Extrait de la collection*, BE9 859, pp.1-17.

Larkin B. K., Churchill S. W., December 1959, "Heat Transfer by Radiation Through Porous Insulations", *American Institute of Chemical Engineers Journal*, Vol. 4, No. 5, pp. 467-474.

Lee S. C., 1986, "Radiative Transfer through A Fibrous Medium: Allowance for Fiber Orientation", J. Quant. Spectrosc. Radiat. Transfer, Vol. 36, No. 3, pp. 253-263.

Lee S. C., 1988, "Radiational Heat-Transfer Model for Fibers Oriented Parallel to Diffuse Boundaries", J. Thermophysics, Vol. 2, No. 4.

Lee, S. C., 1989, "Effect of Fiber Orientation on Thermal Radiation in Fibrous Media", Int. J. Heat Mass Transfer, Vol. 32, No. 2, pp. 311-319.

Lee S. C., 1990, "Scattering Phase Function for Fibrous Media", Int. J. Heat Mass Transfer. Vol. 33, No. 10, pp. 2183-2190.

Lee S. C., Cunnington G. R., 1998, "Fiber Orientation Effect on Radiative Heat Transfer through Fiber Composites", proc. 7th AIAA/ASME Joint Thermophysics and Heat Transfer Conf., Albuquerque, NM, pp. 1-9.

Lee S. C., Cunnington G. R., 1998, "Heat Transfer in Fibrous Insulation: Comparison of Theory and Experiment", Journal of Thermophysics and Heat Transfer, Vol. 12, pp. 297-303.

Lee, S. C., Cunnington, G. R., 2000, "Conduction and Radiation Heat Transfer in High-porosity Fiber Thermal Insulation", *Journal of Thermophysics and Heat Transfer*, vol. 14, no.2, pp.121-136.

Matthews L. K., Viskanta R., and Incropera F. P., 1984, "Development of Inverse Methods for Determining Thermophysical and Radiative Properties of High Temperature Fibrous Materials", *International Journal of Heat and Mass Transfer*, Vol. 2, No.1, pp. 78-81.

Matthews L. K., Viskanta R., and Incropera F. P., 1985, "Combined Conduction and Radiation Heat Transfer in Porous Materials Heated by Intense Solar Radiation", *Journal of Solar Energy*, Vol. 107, pp. 29-34s.

McElroy D.L, Graves R. S., Yarbrough D.W. and Tong T. W. , 1986, "Non-Steady-State Behavior of Thermal Insulations", J. THERMAL INSULATION, Vol. 9, pp 236-249.

Milandri, A., Asllanaj, F., and Jeandel, G. , 2002, "Determination of Radiative Properties of Fibrous Media by an Inverse Method- Comparison with the Mie Theory", *Journal of Quantitative Spectroscopy and Radiative Transfer*, vol. 74, no. 5, pp. 637-653.

Milandri A., Asllanaj F., Jeandel G. , 2002, "Determination of Radiative Properties of Fibrous Media by an Inverse Method- Comparison with the MIE Theory", *Journal of Quantitative Spectroscopy and Radiative Transfer, 74, 5,* pp. *637-653.*

Milandri A., Asllanaj F., Jeandel G., and Roche, J. R. , 2002, "Heat transfer by Radiation and Conduction in Fibrous Media without Axial Symmetry", *Journal of Quantitative spectroscopy and Radiative Transfer*, vol. 74, pp. 585-603.

Milandri, A., Asllanaj, F., Jeandel, G., Roche, J. R., and Bugnon S., 1999, "Transfert de Chaleur Couple Par Rayonnement et Conduction en Regime Permanent dans des Milieux Fibreux Soumis a des Conditions de Flux et en l'absence de Symetrie Azimutale", IVe Colloque interuniversitaire franco-québécois Thermique des systèmes à température modérée.

Nicolau V. P., Raynard M., and Sacadura J. F., 1994 "Spectral Radiative Properties Identification of Fiber Insulating Materials", *International Journal of Heat and Mass Transfer*, Vol. 37, Supplement 1, pp. 311-324.

Nisipeanu, E., and Jones, P. D. , 2003, "Monte Carlo Simulation of Radiative Heat Transfer in Coarse Fibrous Media", *Journal of Heat Transfer*, vol. 125, no. 4, pp. 748-752.

Papadopoulos A. M., 2003, "State of the art in thermal insulation materials and aims for future developments", *Energy and Buildings*, 37, pp. 77–86.

Petrov V. A., 1997, "Combined Radiation and Conduction Heat Transfer in High Temperature Fiber Thermal Insulation", *International Journal of Heat and Mass Transfers,* Vol. 40, No. 9, pp. 2241-2247.

Rish J. W., Roux J. A. , January 1987 , "Heat Transfer Analysis of Fiberglass Insulations With and Without Foil Radiant Barriers", *J. Thermophysics*, Vol. 1, No. 1.

Roux J. A., Smith A. M., August 1977, "Combined conductive and Radiative Heat Transfer in a Absorbing and Scattering Medium", ASME Heat Transfer Conference.

Roux, J.A., 2003, "Radiative properties of high and low density fiberglass insulation in the 4-38.5 µm wavelength region", J. of Thermal Env. & Bldg. Sci. 27(2), pp. 135-149.

Saatdjian E., Demars Y., Klarsfeld S., Buck Y. , 1983, "Effects of Binder Decomposition on High- Temperature Performance of Mineral Wool Insulation", *Thermal Insulation, Materials and systems for Energy Conservation in the 80s*, ASTM STP 789, F. A. Govan, D. M. Greason, and J. D. McAllister, Eds., American Society for Testing and Materials, Philadelphia, pp. 757-777.

Saboonchi A., Sutton W. H., and Love T. J., 1987, "Direct Determination of Gray Participating Thermal Radiation Properties of Insulating Materials", *J. Thermophysics*, Vo. 2, No. 2, pp. 97-103.

Shirtliffe C. J., 1981, "Effect of Thickness on the Thermal Properties of Thick Specimen of Low-Density Thermal Insulation", *National Research Council Canada, No. 966.*

Tong, T. W., Tien, C. L., 1980, "Analytical models for thermal radiation in fibrous insulations", *Journal of Thermal Insulation and Building Envelopes*, vol. 4, pp. 27-44.

Toor J. S., and Viskanta R., 1968, "A numerical Experiment of Radiant Heat Interchange by the Monte Carlo Method", *Int. J. Heat Mass Transfer*, Vo. 11, pp. 883-897.

Yeh H. Y., Roux J. A., 1990, "Transient coupled Conduction and Radiation Heat Transfer through Ceiling Fiberglass/Gypsum Board Composite", Insulation Materials,

Testing and Applications, ASTM STP 1030, D. L. McElroy and J. F. Kimpflen, Eds. American Society for Testing and Materials, Philadelphia, pp.545-560.

Yuen W. W., Takara E., and Cunnigton G., 2003, "Combined Conductive/Radiative Heat Transfer in High Porosity Fibrous Insulation Materials: Theory and Experiment", *Proc. 6th ASME-JSME Thermal Engineering Joint Conference*, Hawaii, USA.

Zeng, S. Q., Hunt, A. J. Greif, R. and Cao, W. , 1995, "Approximate Formulation for Coupled Conduction and radiation Through a Medium with Arbitrary Optical Thickness", *Journal of Heat Transfer*, vol. 117, pp. 797-799.